T0328682

Cambridge Elements

Elements in the Philosophy of Mathematics
edited by
Penelope Rush
University of Tasmania
Stewart Shapiro
The Ohio State University

NUMBER CONCEPTS

An Interdisciplinary Inquiry

Richard Samuels
The Ohio State University

Eric Snyder
Ashoka University

CAMBRIDGE
UNIVERSITY PRESS

Shaftesbury Road, Cambridge CB2 8EA, United Kingdom

One Liberty Plaza, 20th Floor, New York, NY 10006, USA

477 Williamstown Road, Port Melbourne, VIC 3207, Australia

314–321, 3rd Floor, Plot 3, Splendor Forum, Jasola District Centre, New Delhi – 110025, India

103 Penang Road, #05–06/07, Visioncrest Commercial, Singapore 238467

Cambridge University Press is part of Cambridge University Press & Assessment, a department of the University of Cambridge.

We share the University's mission to contribute to society through the pursuit of education, learning and research at the highest international levels of excellence.

www.cambridge.org
Information on this title: www.cambridge.org/9781009462532

DOI: 10.1017/9781009052337

First published 2024

A catalogue record for this publication is available from the British Library.

ISBN 978-1-009-46253-2 Hardback
ISBN 978-1-009-05553-6 Paperback
ISSN 2399-2883 (online)
ISSN 2514-3808 (print)

Cambridge University Press & Assessment has no responsibility for the persistence or accuracy of URLs for external or third-party internet websites referred to in this publication and does not guarantee that any content on such websites is, or will remain, accurate or appropriate.

Number Concepts

An Interdisciplinary Inquiry

Elements in the Philosophy of Mathematics

DOI: 10.1017/9781009052337
First published online: January 2024

Richard Samuels
The Ohio State University

Eric Snyder
Ashoka University

Author for correspondence: Eric Snyder, eric.snyder@ashoka.edu.in; Richard Samuels, samuels.58@osu.edu

Abstract: This Element, written for researchers and students in philosophy and the behavioral sciences, reviews and critically assesses extant work on number concepts in developmental psychology and cognitive science. It has four main aims. First, it characterizes the core commitments of mainstream number cognition research, including the commitment to representationalism, the hypothesis that there exist certain number-specific cognitive systems, and the key milestones in the development of number cognition. Second, it provides a taxonomy of influential views within mainstream number cognition research, along with the central challenges these views face. Third, it identifies and critically assesses a series of core philosophical assumptions often adopted by number cognition researchers. Finally, the Element articulates and defends a novel version of pluralism about number concepts.

Keywords: number concepts, number cognition, representationalism, core systems, conceptual pluralism

ISBNs: 9781009462532 (HB), 9781009055536 (PB), 9781009052337 (OC)
ISSNs: 2399-2883 (online), 2514-3808 (print)

Contents

1 Introduction

This Element is principally concerned with the conceptual building blocks of numerical thoughts. These include, but needn't be limited to, thoughts ascribing cardinalities to collections, such as the thought expressed by (1a), those ascribing arithmetic properties to numbers, such as that expressed by (1b), and those locating the ordinal position of something among a series of things, such as the thought expressed by (1c).

(1)   a. The Elmos on the table are two in number.

      b. Two is an even number.

      c. Mary is contestant number two.

Generally, as constituents of thoughts, concepts are widely assumed by philosophers and psychologists to be essential to a range of cognitive capacities, most obviously categorization, reasoning, language comprehension, and decision-making. For example, *understanding* (1a–c) requires the possession of concepts; having the thoughts expressed by each sentence permits the *categorization* of some thing(s) as being a certain way – as being two in number, an even number, or second among the contestants; and each thought clearly permits *reasoning*, e.g. that there is more than one Elmo on the table, that some number is even, or that Mary is a contestant. Having such thoughts requires the possession of concepts which are broadly numerical in nature, or as we call them, **number concepts**.[1]

In writing this Element, we have a pair of central aims. The first is to summarize and critically evaluate empirical research on number cognition of the sort pursued in developmental psychology and cognitive science – which we will call **number cognition research** (NCR) – and to do so in a manner that is helpful to philosophers with little or no background in that research. The second aim is to highlight how research from disparate fields – such as, the philosophy of mathematics, developmental psychology, and linguistics – can be mutually informing, in previously unappreciated ways. In particular, by highlighting theoretical assumptions underlying much NCR, and by bringing insights from the philosophy of mathematics and linguistics to bear on these assumptions, we hope to spark further interdisciplinary dialog.

Here, we outline how the Element attempts to achieve this second aim, by considering why number cognition researchers (NCRs) should care about the

---

[1]   A note on the conventions employed. We use boldface to indicate technical terms, and typically reserve italics for emphasis. Single quotes are used when mentioning expressions, such as the word 'dog', and double quotes are used for direct quotation. Finally, we use capital letters to designate concepts, as with the concept DOG.

philosophy of mathematics, and why philosophers of mathematics should care about NCR. Along the way, we highlight key issues raised throughout the Element, while indicating the relevant sections.

1.1 Why Care about Philosophy of Mathematics?

One reason NCRs should care about the philosophy of mathematics is that the sorts of foundational issues studied within this area of philosophy routinely inform NCR. Specifically, as we'll see, NCRs often adopt controversial philosophical assumptions regarding such issues as what natural numbers are, and how they are represented in language and thought.

To be clear, there is nothing wrong with importing foundational assumptions into NCR. Indeed, it may be unavoidable. Nevertheless, we want to highlight five points regarding the manner in which such assumptions figure within NCR. First, typically these assumptions are only a small sample of the available options. Second, and despite this, justification for the preferred options are rarely provided. Third, the choices made often have an important effect on how NCR is conducted. Fourth, there are often independent problems with the options chosen. Finally, sometimes adopting a neglected alternative would both avoid these problems and cast the empirical issues in an importantly different light.

To illustrate, consider the predominant metaphysical assumption about natural numbers within NCR: **the Frege–Russell characterization**. On this view, the naturals are, loosely speaking, collections of sets having the same number of members. For example, the number two is identified with the collection of all two-membered sets, the number three with the collection of all three-membered sets, and so on. As we'll see in Section 4.2.3, despite its undeniable influence both within and outside of philosophy, the adoption of the Frege–Russell characterization within NCR has not only exerted a significant effect on how research is conducted, it has also led to certain theoretical problems which have not been widely appreciated. Furthermore, there are available alternatives which, if adopted, would both avoid these problems and have important implications for a range of relevant empirical issues.

First, note that the Frege–Russell characterization is only one version of several kinds of characterizations of the naturals available. Specifically, it is a *cardinal* characterization – one characterizing the naturals as answering 'how many'-questions. However, not only are there other cardinal characterizations available, there are also different *kinds* of characterizations altogether, namely ordinal and structuralist characterizations (Section 4.1).

Second, although the Frege–Russell characterization is very frequently adopted by NCRs, its adoption is rarely, if ever, explicitly justified. Indeed, as Rips et al. (2008, p. 625) observe, despite the availability of alternatives,

> Nearly all cognitive research on the development of number concepts rests on the idea that such concepts depend on enumerating objects . . . Psychologists seem to have been influenced . . . by the conception of natural numbers as sets of all equinumerous sets of objects (see Frege 1884/1974; Russell 1920).[2]

For example, after querying where number concepts come from and how they develop, vanMarle (2018, p. 131) immediately provides an answer to the question of what numbers are: "As described by Frege (Frege, 1884/1980), formally, numbers are a special kind of sets."[3]

Third, and despite this, adopting the Frege–Russell characterization has important implications for central issues in NCR, including the following:

- *Denotations of Count Words*: **Count words** – words used for counting, such as 'one', 'two', 'three', and so on – are assumed to denote natural numbers, as collections of sets (Section 4.2.3). For example, Bloom and Wynn (1997, p. 512) write: "[In *two black cats*], *two* is a predicate that applies to the set of cats. As Frege (1893/1980) has argued, numbers are predicates of sets of individuals." The presumption appears to be that because 'two' predicates twoness of sets, the number two must be the extension of that predicate. However, in addition to being a dubious interpretation of Frege (Benacerraf [1965]), the assumed analysis is not the only – or even the most prominent – analysis of count words available (Section 4.2.3).
- *Representational States of Core Systems*: Certain so-called "core systems" (Section 2.2), such as the Approximate Number System (ANS), the Small Number System (SNS), and the language faculty, are assumed to deploy representations of sets. For example, regarding the ANS, Susan Carey (2009a, p. 136) writes: "The analog magnitude number representations …require representations of sets." Similarly, concerning the SNS, Carey (Carey, 2009a, p. 150) writes: "Parallel individuation, like analog magnitude number representation, depends on representations of sets and supports quantitative computations over sets." Finally, regarding natural language semantics, Carey (Carey, 2009a, pp. 320–321) writes: "Semantic treatments of quantifiers require the abstract concepts individual and set." Again, however, this

---

[2] Within contemporary ZF set theory, natural numbers are not *sets* (of equinumerous sets), of course, since no such sets exist. Rather, they are *classes*.

[3] For Frege, natural numbers are extensions of concepts (Frege [1884]) or courses of values (Frege [1903]). If we interpret these as sets, then Frege's account is inconsistent, of course, due to Russell's paradox.

is not the only, let alone the most plausible, way of describing what these systems represent (Section 4.2.3).

- *Representational Capacities*: It is sometimes assumed that because number concepts denote sets, acquiring such concepts requires representational capacities implied by certain foundational programs, such as Zaremlo–Fraenkel set theory or Frege arithmetic (Section 4.1). Quoting Carey (2009b, pp. 1–2):

> ([C]haracterizing the logical prerequisites for natural number) leads to analyses like those that attempt to derive the Peano–Dedekind axioms from Zermelo–Fraenkel set theory or Frege's proof that attempts to derive these axioms from second order logic and the principle that if two sets can be put in 1–1 correspondence they have the same cardinal value. Such analyses seek to uncover the structure of the concept of natural number, and certainly involve representational capacities drawn upon in mature mathematical thought.

However, these are not the only foundational programs capable of deriving the Dedekind-Peano axioms (Snyder et al. [2019]). Also, if acquiring the relevant concepts involves an induction from perceptual inputs, then acquiring such concepts would appear to require a capacity to perceive abstract objects. In contrast, available alternatives would require no such capacity (Section 4.2.3).

Fourth, despite its influence, there are significant problems with adopting the Frege–Russell characterization within the context of NCR. For example, in addition to potentially requiring perception of abstracta, when combined with independently plausible assumptions concerning the semantic functions of number words, the Frege–Russell characterization makes numerous false semantic predictions (Section 4.2.3).

Finally, we suggest in Section 4.3 that replacing the Frege–Russell characterization with other foundational commitments both avoids some of the problems just alluded to and casts certain core issues about number cognition in an importantly different light. For example, once the Frege–Russell characterization is relinquished, we maintain that it is plausible to view the available semantic evidence as supporting a novel kind of **conceptual pluralism** about number concepts – one in which adults possess multiple distinct, but semantically related, concepts for each number word in their lexicon.

1.2　Why Care about Number Cognition Research?

Although this Element focuses primarily on the relevance of the philosophy of mathematics for NCR, we think it important to comment briefly on why philosophers should also care about NCR, especially since many contemporary philosophers of mathematics may view such research as irrelevant to their own concerns, for at least three reasons.

First, thanks largely to the influence of Frege, much philosophy of mathematics is **anti-psychologistic**. Specifically, Frege (1884, p. xviii) expresses an attitude which, we suspect, is still largely prevalent today: "psychology should not imagine that it can contribute anything whatever to the foundation of arithmetic." Given this sentiment, we should not expect psychology to play a role in the sorts of foundational matters that frequently concern philosophers of mathematics.

Second, unlike NCR, much of the research within philosophy of mathematics is *revisionist* in spirit. Strawson (2002) famously distinguishes two methodological orientations toward metaphysics. Roughly, whereas **descriptive metaphysics** is concerned with what the structure of reality would be like if it were accurately represented by the conceptual scheme we *actually* have, **revisionary metaphysics** is concerned with what the structure of reality would be like if it were accurately represented by the conceptual scheme we *ought* to have. At least since Frege, and his attempt to develop an ideal language suitable for science, much philosophy of mathematics has been rather more concerned with the latter. So, since NCR is concerned with how we *in fact* think, the significance of such research for the philosophy of mathematics is once more attenuated.

Finally, many philosophers of mathematics view themselves as principally concerned with mathematics as a *scientific discipline* – one that's populated by specialists, typically engaged in highly creative forms of research that are beyond the comprehension of all but a highly select group of other experts. Far less concern, overall, is given to elementary arithmetic, and even less to our mundane thoughts about counting Elmos or balancing checkbooks. Thus, it may seem unobvious why philosophers should concern themselves with childhood conceptual development, let alone with numerical capacities we share with nonhuman organisms. Indeed, from this vantage, it may seem about as relevant as it would be for, say, philosophers of physics to study the childhood development of folk physics, or for philosophers of literature to study how children learn to read. In all such cases, the gap between the quotidian capacities studied by psychologists and the forms of expertise that philosophers reflect upon may appear so great as to render the former irrelevant to the respective philosophical projects.

Nevertheless, there are at least three good reasons philosophers of mathematics should care about NCR.[4] First, empirical claims of the sort which

---

[4] For additional potential points of contact, see de Cruz (2016).

concern NCR are already relatively commonplace in developing and defending influential positions within the philosophy of mathematics. Thus, insofar as philosophers rely on such claims, they should ensure that their commitments are actually supported by our best NCR.

To illustrate, consider the **neo-logicism** of Hale and Wright (2001). One of their central metaphysical commitments is that natural numbers are finite cardinalities characterized by **Hume's Principle** (HP), where 'F' and 'G' range over predicates, '$\#$' is an operator mapping predicate extensions to cardinal numbers, and '$\sim$' is the relation of **equinumerosity** holding between F and G if each F can be mapped to a unique G, and vice versa.

(HP) $\forall F, G.\ \#F = \#G \leftrightarrow F \sim G$

As Hale and Wright are aware, this is not the only characterization of the naturals available (Section 4.1). Consequently, they seek to justify (HP) as the *correct* characterization of the naturals, to the exclusion of alternatives. To this end, they appeal to a principle known as **Frege's Constraint**: roughly, the correct formal characterization of the naturals ought to directly reflect their primary empirical application [Snyder et al. (2018a)]. What's more, they argue that the primary empirical application of the naturals is *counting*. For Wright (2000, p. 327) in particular, the metaphysical thesis is justified partly in virtue of how children acquire number concepts.

> It seems clear that one kind of epistemic access to …simple truths of arithmetic proceeds precisely through their applications. Someone can – and our children surely typically do – first learn the concepts of elementary arithmetic by a grounding in their simple empirical applications and then, on the basis of the understanding thereby acquired, advance to an a priori recognition of simple arithmetic truths.

Thus, "there is a kind of a priori arithmetical knowledge which flows from an antecedent understanding of the way that arithmetical concepts are applied" – namely through counting. Yet whether children typically acquire arithmetic knowledge only after learning how to count is, of course, an *empirical* claim, one beholden to the developmental facts. Fortunately for Wright, those facts do appear to support this contention (Section 2.2).[5]

As a second example, consider **mathematical empiricism**, as defended, e.g. by Kim (1981), Maddy (1990), and Yi (2018). As usually formulated, empiricism is principally an empirically motivated response to Benacerraf's (1973)

---

[5] Though see Linnebo (2009). Also, whether this actually *justifies* Frege's Constraint, or indeed whether it establishes counting as the primary empirical application of the naturals, is an altogether different matter, one which we are highly skeptical of [Snyder et al. (2018a, 2019)].

Access Problem: roughly, how is it possible to have mathematical knowledge if such knowledge is about abstract mathematical objects? Following J. S. Mill, modern mathematical empiricists propose that our knowledge of simple arithmetic truths, e.g. that $3 + 2 = 5$, depends crucially on our perceiving the cardinalities of collections of concrete entities. Two theses are thus characteristic of such empiricism. The first is

The Identity Thesis: Natural numbers are cardinalities.

By **cardinality**, we mean the sort of thing answering 'how many'-questions. Different versions of empiricism offer different accounts of what cardinalities are – e.g. properties of sets [Kim (1981); Maddy (1990)]. Still, all maintain

The Perceptual Thesis: Cardinalities are perceivable.

Empiricists often appeal to NCR in attempting to justify this thesis. Specifically, while Kim (1981) and Maddy (1990) both appeal to **subitizing** (Section 2.2), and thus presumably some version of the SNS, Yi (2018) appeals instead to the ANS. On the presumption that the SNS and ANS are *perceptual* systems representing cardinalities, empiricists thus seek to explain how our knowledge of basic arithmetic may be grounded in our perception of cardinality, and thus *number*.

Evidently, whether this defense of mathematical empiricism is successful depends on empirical evidence for the existence of these systems, as well as the contents of the representations such systems deploy. As we will see, while most Mainstream NCRs appeal to one or more of these systems in accounting for the possession of number concepts, not all theories appeal to the same systems, and not all agree that the representations deployed by such systems suffice to possess number concepts (Section 3).

A second potential reason philosophers should care about NCR is closely related to the Access Problem. Though there are different formulations of the Access Problem, by far the most familiar are epistemic in character. Specifically, whereas Benacerraf's (1973) original formulation concerns how it is possible to possess the sorts of mathematical *knowledge* we seem to have, Field's (1989) reformulation concerns how we can explain the *reliability* of our mathematical beliefs. Yet as Hart (1991) notes, a presupposition of such problems – and a large part of what makes them difficult to address – is the challenge of explaining how humans can have thoughts *about* numbers in the first place. However, this has been a central issue for NCR since its inception, with Walter McCulloch's well-known paper "What is a number, that a man may know it, and a man, that he may know a number?"

Though the details of McCulloch's paper needn't detain us, the challenge of explaining how we can have thoughts about numbers decomposes into two subproblems. The first is a challenge for theories of mental representation:

The Content Problem: How can thoughts have numerical content, if such a content is partially constituted by abstracta?

Consider the thought that two is an even number, for example. Prima facie, this thought is partially constituted by the concept TWO, which, like the numeral 'two' in (2), presumably denotes a number.

(2)   Two is an even number.

The question is *how* could this thought have such a content, given what TWO denotes? What's needed, apparently, is a theory of **content determination** which allows for such a possibility. To see why, consider the following passage from Carey (2009a, p. 5):

> I assume that there are many components to the processes that determine conceptual content, and that these fall into two broad classes: (1) causal mechanisms that connect a mental representation to the entities in the world in its extension, and (2) computational processes internal to the mind that determine how the representation functions in thought. In broad strokes, then, I assume a dual theory of conceptual content (Block, 1986, 1987).

Here Carey assumes a *causal* condition on content determination whereby a representation ϕ denotes X only if there are causal mechanisms linking ϕ to X. Yet it seems impossible that thoughts about numbers could, in principle, be *causally* related to numbers, assuming numbers are abstract.

The second subproblem would have been recognizable to Plato:

The Acquisition Problem: How is it so much as possible for humans to *acquire* number concepts?

Obviously, an answer here requires an adequate answer to the Content Problem. But it also requires the specification of processes that permit humans to acquire representations with the relevant contents. One may fruitfully view large parts of NCR as dedicated to specifying such processes. As we'll see, there is broad agreement regarding many of the relevant ingredients – e.g. cognitive systems – involved in acquiring number concepts (Section 2.3), and also the sequence of salient developmental milestones (Section 2.4). Yet there is substantial disagreements concerning the details. Section 3 contains a sustained discussion of such disagreements; and Section 4 contains an assessment of central logico-mathematical assumptions that dominate the field.

We are mindful that while The Content Problem is widely presumed to be a philosophical issue, many philosophers of mathematics will view the Acquisition Problem as falling outside their discipline. Fregean scruples notwithstanding, however, the challenge of explaining how such concepts can be acquired is a topic of enduring philosophical interest, at least since Plato. Moreover, as we hope to make clear, the Acquisition Problem is impacted by precisely the sorts of issues that *do* figure centrally in philosophy of mathematics. Such considerations, we suggest, provide ample grounds for the continued attention of philosophers.

1.3 Ambitions and Scope of the Present Element

The primary purpose of this Element is to critically review NCR in a manner that is helpful to both philosophers and NCRs. However, we do not aim to provide a detailed exposition of the myriad experiments, data, models, and hypotheses that pervade NCR. Instead, we focus on the broadly philosophical assumptions that underwrite such research. For one thing, there are already multiple volumes, written or edited by first-rate cognitive scientists, which provide excellent surveys of such empirical details (see, e.g. Kadosh and Dowker [2015]). For another, our talents, such as they are, are better suited to wrangling the many, often underexplored, theoretical underpinnings of NCR.

We should also stress that while much of our discussion is critical, the criticisms presented are intended to be constructive, not dismissive. This Element is part of a larger, long-term project looking at those empirical sciences most clearly relevant to how we talk and think about number, and attempting to discern how, if at all, they may fruitfully inform foundational issues within the philosophy of mathematics, and vice versa. Ultimately, then, our hope is that this Element contributes to a more fruitful interdisciplinary dialog between philosophers, NCRs, and linguists.

The remainder of the Element is organized into three sections. Section 2 cordons off a large group of influential theories of number cognition, which we call "the Mainstream." It does so by identifying certain commitments characteristic of much NCR, including representationalism about cognition (Section 2.1), and representationalism about concepts in particular (Section 2.2), the existence of a pair of so-called "core systems" (Section 2.3), and a broad consensus regarding some important developmental milestones for number cognition (Section 2.4). In Section 3, we propose a taxonomy of Mainstream views, based roughly on which factors they deem central to the acquisition of number concepts, and critically review some influential theories in each category. These include what we call ANS-dominant models (Section 3.2), nativist models (Section 3.3), hybrid models (Section 3.4), and SNS-dominant models (Section 3.5). Finally, in Section 4, we isolate and critically assess certain philosophically relevant theoretical assumptions underlying Mainstream NCR. These include methodological assumptions concerning the relationship between NCR and foundational theorizing (Section 4.1), as well as various metaphysical positions concerning natural numbers, including social constructionism (Section 4.2.1), term formalism (Section 4.2.2), and the Frege–Russell characterization (Section 4.2.3). We end the Element by distinguishing two kinds of conceptual pluralism, and suggesting that both are plausible in light of broadly empirical considerations (Section 4.3).

2 The Mainstream

At least since Plato, mathematical cognition in general, and numerical cognition in particular, has been a topic of central interest for philosophers. However, it is only in the past four decades or so that it has become the subject of sustained empirical research – most clearly in developmental psychology, cognitive psychology, and cognitive neuroscience.

Although the field is replete with substantive, ongoing disagreements, much contemporary number cognition research (NCR) cleaves to an array of empirical and theoretical commitments, which may be thought of as constituting a kind of Mainstream view of number cognition. In saying this, we do not intend to suggest that all these commitments are explicitly endorsed by all, or even most, contemporary NCRs. Rather, our talk of "the Mainstream" should be viewed as a useful idealization – a kind of conceptual "central tendency" implicit in much contemporary research. And like many idealizations, we take our talk of the Mainstream to be warranted to the extent that it proves useful in systematizing and aiding understanding.

The task of the present section is to articulate, not to defend or to critically evaluate, the constellation of commitments which constitute the Mainstream. Though we will, of course, present common reasons for endorsing a given commitment, and refer the reader to existing relevant research, the task of critically evaluating Mainstream commitments is deferred until later.

For heuristic purposes, we divide Mainstream commitments into two sorts: general commitments that are pervasive across the cognitive sciences, and those specifically concerning the domain of number cognition. The first are inherited from cognitive science as a broad, ongoing concern and applied specifically to the domain of number cognition. For our purposes, the most significant of these concern *mental representation*. Specifically, as we discuss below, in order to understand Mainstream NCR is it important to appreciate (i) the widespread commitment to representationalism as a general thesis about cognition; (ii) the presumed explanatory function of representations in cognitive explanations; and (iii) the widespread commitment to the view that concepts are a species of mental representation.

The second, number-specific variety of commitment can be divided into two further sorts. The first are broadly empirical commitments supported by decades of NCR. Three such commitments will be especially significant in the following discussion, both because they constitute a kind of broad empirical consensus, and also because they structure ongoing research efforts:

The Core Systems Hypothesis: There are at least two evolved, cognitive systems – sometimes called the 'small number system' and the 'approximate number system' – that humans share with many other animals, and which underwrite processes which address numerical tasks, broadly understood.

The Developmental Timetable: Up until approximately 4 years of age, infants and children tend to manifest a robust, stage-like pattern in the development of numerical abilities.

The Primacy of Counting: Learning to count plays a central role in the manifestation of early emerging numerical abilities.

In addition to these empirical commitments, a second kind of number-specific commitment concerns what the *big questions* – the central explanatory *targets* – for a science of number cognition should be. Interestingly, few researchers directly aim to explain the sorts of advanced numerical abilities manifested by professional mathematicians. Instead the aim is to understand more basic, mundane – or *quotidian* – numerical abilities routinely manifested in industrial societies from, say, four years onwards: most centrally, counting and simple arithmetic. In this regard, the science of number cognition is like the science of language, which does not directly aim to explain exceptional linguistic feats – of Keats, Flaubert, or Angelou, for example – but instead to understand the sorts of linguistic competence possessed by almost any human being. As we'll see, perhaps the central explanatory target for contemporary NCR concerns how young children come to manifest simple quotidian numerical abilities – especially those indicative of the possession of concepts denoting natural numbers (ONE, TWO, etc.), tacit knowledge of numerical principles, such as successor, and the concept NATURAL NUMBER as such.

2.1 Representationalism

Among the most widely shared assumptions in all of cognitive science is what's sometimes called **the Representational Theory of Mind** (RTM), or just **representationalism** [Pinker (2007), Carey (2009a)]. Representationalism has, of course, been the focus of extensive critical discussion over the past half century, and is the subject of some excellent book-length treatments [Fodor (1990); Ramsey (2007); Shea (2018)]. Hence, we won't belabor the details here. Still, it is important to be clear on various aspects of contemporary representationalism, if we are to understand extant NCR.

Most generally, representationalism is the thesis that the human mind is an information-using system, and that our cognitive capacities can be understood in representational terms [Egan (2012)]. The core idea is that psychological processes, such as thinking, perceiving, language comprehension, and learning, are causal processes involving information-bearing structures or vehicles – **mental representations** – which have appropriate *contents* or *meanings* [Shea (2018)]. The models and explanations provided by cognitive scientists quite generally, and NCRs in particular, most often assume the existence of such processes.

Many processes studied by cognitive scientists do not correspond in any straightforward manner to our commonsense ways of classifying cognitive processes. For example, as we'll see in Section 2.3, cognitive scientists have extensively studied processes for the representation of approximate magnitudes, and for perceptually tracking small numbers of objects. However, if we restrict our attention to those cognitive processes philosophers most typically discuss – those involving propositional attitudes, such as beliefs and desires – then representationalism may be usefully presented as follows, quoting Margolis and Laurence (2022).

> According to RTM, thinking occurs in an internal system of representation. Beliefs and desires and other propositional attitudes enter into mental processes as internal symbols. For example, Sue might believe that Dave is taller than Cathy, and also believe that Cathy is taller than Ben, and together these may cause Sue to believe that Dave is taller than Ben. Her beliefs would be constituted by mental representations that are about Dave, Cathy and Ben and their relative heights. What makes these beliefs, as opposed to desires or other psychological states, is that the symbols have the characteristic causal-functional role of beliefs.

In practice, advocates of representationalism in cognitive science tend to adopt a suite of assumptions about the representations which, by hypothesis, underwrite cognition.

First, though the assumption won't figure prominently in the coming discussion, cognitive scientists usually adopt a computational model of cognition, and so characterize the causal-functional roles of representations in computational terms Harnish (2000).

Second, for explanatory reasons, almost all advocates of representationalism take mental representations to have internal structure. Accordingly, those representations figuring in beliefs, for instance, are often presumed to be composed of more basic representations.

Third, like all representations, mental representations are presumed to have semantic properties – meanings, truth-conditions, denotations, or more simply "contents" when distinctions between these are unimportant. Although there is considerable disagreement regarding which specific semantic properties mental representations possess – whether, e.g. they possess properties akin to Fregean senses – they are widely assumed to have denotational or referential contents. For example, it is routinely assumed that some mental representations function like **singular terms**, i.e. expressions whose semantic function is to refer to entities, while other mental representations function like predicative expressions, in that they purport to denote properties or collections of objects.

Fourth, in addition to their semantic properties, mental representations are assumed to possess nonsemantic properties which explain why representations with the same contents are treated in the same ways by cognitive processes [Shea (2018)].

Fifth, it is important to appreciate that within cognitive science, mental representations are posited largely in the service of providing various sorts of *explanation*. Specifically, a primary motivation for positing mental representations is to causally explain empirically attested patterns in behavior and cognition. It is ultimately on its adequacy as an approach to the provision of such explanations that representationalism stands or falls.

One style of explanation, prominent in developmental psychology, will be especially pertinent to our discussion in later sections, and consists in explaining cognitive-behavioral differences by positing *representational* differences. In particular, one seeks to explain observed contrasts in performance between members of developmentally distinct populations – say, two year-olds and four year-olds – by positing a representational resource, such as a concept or body of information, that one population is assumed to lack and the other to possess. As we'll see, such explanations are routinely invoked by NCRs to explain differences in numerical abilities at different times in development.

Finally, it is important to appreciate that, in providing such explanations, cognitive scientists routinely assume a familiar inventory of representational states, including those marked by the following distinctions:

- *Symbolic and Analog Representations*: It is widely supposed that some representations may be **symbolic**, or language-like, in that they have a combinatorial syntax and a compositional semantics [Fodor (1975)]. However, it is also widely assumed that there may be **analog representations** of various sorts (Section 2.3). How best to characterize these is, in our view, not at all clear, and remains a point of ongoing debate [Beck (2019); Lee et al. (2022)]. However, presumed paradigmatic (noncognitive) examples include mercury thermometers, hand-clocks, and line drawings [Lee et al. (2022)].

- *Explicit and Implicit Representational States*: At least since Chomsky's early work, it has been widely supposed that there are both explicit and implicit representational states. Roughly put, a representation is explicit when a statement describing its content may be elicited by suitable enquiry or prompting [Dummett (1991b)]. In contrast, when no such statements can be elicited, a representational state is often labeled "implicit" or "tacit." Though there is considerable disagreement regarding how best to characterize implicit representations (see, e.g. [Davies (2015)], what's crucial for the purposes of NCR is that they can participate in causal interactions, even if they are not reportable by the agent.

- *Conceptual and Nonconceptual Representations*: The final distinction is between conceptual and nonconceptual representational states. Paradigmatic instances of the latter include states of early vision and the language faculty, whereas paradigmatic instances of the former include propositional attitudes, e.g. beliefs and judgments [Bermúdez and Cahen (2020)]. As we'll see, much of NCR assumes that number concept acquisition depends on a prior possession of representational capacities that are nonconceptual in character.

2.2 Representationalism about Concepts

Suppose, along with the Mainstream, that some version of representationalism is correct. Further, suppose that we are concerned with concept acquisition. What kind of entity, quite generally, will concepts be? In what follows, we sketch some commonplace assumptions in cognitive science, especially developmental psychology [Carey (2009a)].

First, concepts are standardly viewed as the building blocks, or constituents, of thoughts and other propositional attitudes [Margolis and Laurence (2022)]. According to the Mainstream, thoughts are mental representations, and concepts qua constituents of thoughts, are themselves mental representations. Here, 'thought' is being used not merely to indicate occurrent, conscious episodes, but also unconscious thoughts, as well as various sorts of standing mental states, including long-term memories and mentally represented theories [Carey (2009a)]. By broad consensus, many of these states may never be consciously accessible to the agent.

Second, cognitive scientists quite broadly, and NCRs in particular, tend to suppose that concepts, as a species of mental representation, play characteristic cognitive roles. Specifically, concepts are implicated in "higher" cognitive capacities, such as reasoning, language comprehension, categorization, and word learning. The explanation of why concept possession is central to reasoning should be clear: reasoning requires propositional attitudes, such as beliefs or judgements, and concepts are constituents of propositional attitudes. However, the relationship between concept possession and these other capacities may require further explanation.

Start with language comprehension. Plausibly, understanding a sentence is a kind of propositional attitude – roughly, understanding what it says, in context. Thus, to understand a sentence, such as (2), one must have the relevant attitude.

(2)   Two is an even number.

Given representationalism about concepts, such an attitude will have the concept TWO as a constituent, and understanding (2) will require possession of TWO. Thus concepts are implicated in language comprehension.

Similar points apply to categorization judgements and word learning. Presumably, categorization judgements are a species of propositional attitude, namely judgements regarding whether something is a member of a category. In which case, if concepts are constituents of propositional attitudes, then to categorize the number two as even, one must minimally possess the concepts TWO and EVEN.

Finally, consider the case of word learning. Though this phenomenon is complex, it is surely central to learning a word like 'two' that, by doing so, one is in a position to understand at least some sentences in which it figures. But as already indicated, on a representationalist account of concepts, to understand a sentence, such as (2), one must possess the relevant corresponding concepts. In which case, word learning likewise implicates the possession of concepts. Following Fodor and others, we call such concepts **lexical concepts**, since they are concepts corresponding to lexical items – or crudely, words [Fodor (1998)]. Though much ink has been spilt regarding the nature of lexical concepts – e.g. regarding whether they have definitional, prototype, or theory-like structure [Margolis and Laurence (1999)] – they won't figure centrally here. In contrast, issues about the connection between concepts and word learning figure very prominently indeed.

2.3 The Core Systems Hypothesis

The previous commitments are shared quite broadly among cognitive scientists. We now turn to commitments specific to NCR, starting with what we call *the Core System Hypothesis*. This is the view that some rudimentary forms of number cognition rely on **core systems** – cognitive systems that prototypically possess all, or at least most, of the following characteristics [Spelke (2000); Spelke et al. (2022); Carey (2009a)]:

- *Evolved*: They are evolved traits produced by natural selection.
- *Highly Conserved*: They are shared by both human and nonhuman organisms, including nonhuman primates, other mammals, and perhaps also avian species.
- *Domain Specific*: They are **domain specific**, roughly in that they evolved to operate on a restricted range of perceptual inputs.
- *Task Specific*: They are **task specific**, roughly in that they are designed to solve specific kinds of problems.
- *Ontogenetically Precocious*: They typically operate from early infancy onwards.
- *Independent*: Their operation is relatively independent of other cognitive systems.
- *Innate*: They are innate, or at least not acquired via learning processes.

Although there is disagreement regarding how many core systems there are, how they function, and what they represent, at least two core systems are frequently invoked by Mainstream accounts within NCR. The first, often called **the Approximate Number System** (ANS), has been claimed by some influential NCRs to represent large cardinalities, at least approximately. The second, sometimes called **the Small Number System** (SNS), has been claimed by some influential NCRs to represent small numbers of items, including things as diverse as Elmos, sounds, and actions.

Before considering either of these systems, however, we introduce a distinction that will help avoid confusion later on: the distinction between merely *tracking* a property and *representing* it. As we intend it, a system merely tracks a property, in some range of environments, if states of that system *causally covary* with the presence or absence of that property, in those environments. Intuitively, there are many examples in nature where one sort of state tracks or covaries with another without representing it. For example, changes in heart rate causally covary with changes in physical exertion; spots of the relevant sort covary with measles; and smoke covaries with fire. In none of these cases, however, does one state *represent* the other. In our parlance, these are cases of *mere* tracking. Similarly, to say that the SNS merely tracks exact cardinalities up to four (in some range of environments) is to say its states causally covary with the presence of collections having some such cardinality. Without further assumptions, it does not imply that states of the SNS *represent* cardinalities. Of course, some theories of representational content – most obviously causal theories – construe the tracking relation as central to explaining representation [Fodor (1992)]. Yet no plausible theory presumes that tracking alone suffices for representation. Rather, some accounts construe tracking as merely one among a range of conditions that jointly suffice for representation. On other views, however, tracking plays no significant role.

We stress this distinction because while there is widespread consensus within NCR that both the ANS and SNS *track* broadly numerical properties, there's little or no agreement regarding whether they *represent* such properties. Thus, in the case of the SNS, while it's widely supposed that, across a wide array of environments, states of the system track exact differences in cardinality up to 4, there is a remarkable lack of consensus regarding whether such states *represent* these cardinalities. Similarly, with the ANS, while almost everyone supposes that (across many environments) the system's states covary with approximate cardinalities, there's little consensus regarding which, if any, broadly numerical properties the system *represents* (Section 3.2). With this in mind, the forthcoming characterizations of the ANS and SNS seek both to capture the consensus regarding their tracking behavior, and some of the disagreements regarding what they represent. In later sections, we explore issues regarding representation in greater detail.

2.3.1 Approximate Number System

The ANS is a primitive, prelinguistic cognitive system, also sometimes called **the number sense** (Dehaene [2011]). It is distinguished from other cognitive systems by two characteristic features: quoting Clarke and Beck (2021, p. 2), the ANS's "numerical discriminations are imprecise and conform to Weber's Law." Intuitively, the ANS's discriminations are "imprecise" in that instead of tracking, say, *exactly* 20 objects, it instead tracks *approximately* 20 objects.[6] Moreover, it conforms to a psychophysical principle known as **Weber's Law** – its ability to discriminate magnitudes decreases as the ratio between those magnitudes approaches 1:1. Consequently, it is just as easy to distinguish 5 objects from 10 as it is 10 objects from 20, since the ratios between the two pairs are the same. In contrast, it is easier to discriminate 2 from 4 objects than 8 from 10, since the ratios between the two pairs are different, even though the absolute difference is the same. (This is known as the **magnitude effect**.) Similarly, it is easier to distinguish four from eight objects than four from seven, since the ratio is greater. (This is called **the distance effect**.) Figure 1 illustrates typical stimuli testing for such effects.

---

[6] Previous attempts to characterize the manner in which ANS is imprecise often suggest that there is no difference between how the system responds to, say, 20 things and 19 things [Laurence and Margolis (2007); Carey (2009a)]. However, more recent studies suggest that the ANS may not be imprecise in this sense [Sanford and Halberda (2023)].

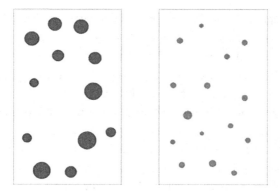

Figure 1 Illustration of typical stimuli for ANS experiments.

As ordinarily construed, the ANS is a kind of **analog magnitude system**, roughly in that states of the system represent **quantities**, and also that states of the ANS involve magnitudes which are approximately proportional to what they represent. According to one influential view, for example, ANS states represent the approximate cardinalities of collections in virtue of instantiating physical magnitudes of some sort – e.g. amounts of electrical charge – roughly proportional to the cardinalities of the collections represented. Carey (2009a) usefully illustrates this feature in terms of a system in which lines of differing lengths represent different cardinalities.

Number	Length
1	-
2	–
3	—
4	—
5	——
6	——
7	——
8	——

The important point is that within this representational system, "number is represented by a physical magnitude that is [merely] *roughly* proportional to the number of individuals in the set being enumerated" (Carey [2009a, p. 118], our emphasis). Precisely how best to characterize the numerical contents of these analog representations remains a point of ongoing debate. For instance, on some views, such states represent real numbers, on others rationals, and on some only approximate cardinalities (Section 3.2.1).

There is plenty of empirical evidence for the existence of the ANS in both humans and nonhuman animals, which we will not review here.[7] And while different models of the ANS exist,[8] the one most frequently cited by Mainstream theories is **the accumulator model** of Meck and Church (1983), the crucial features of which Laurence and Margolis (2005, pp. 218–219) capture with the following metaphor:

> Imagine water being poured into a beaker one cupful at a time and one cupful per item to be enumerated. The resulting water level (a continuous variable) would provide a representation of the numerosity of the set: the higher the water level, the more numerous the set. Moreover, with an additional beaker, the system would have a natural mechanism for comparing the numerosities of different sets. The set whose beaker has the higher water level is the larger set. Similarly, the Accumulator could be augmented to support simple arithmetic operations. Addition could be implemented by having two beakers transfer their contents to a common store. The level in the common store would then represent their sum.

Figure 2, taken from Gallistel and Gelman (2000), illustrates the metaphor, where lines on the beaker represent cardinalities, which map one-to-one with natural numbers.

Of course, no-one is seriously suggesting that the ANS uses beakers of water to represent cardinalities. Instead, the presumption is that electrochemical events and properties – e.g. pulses of electricity and magnitudes of stored electrical charge – are doing the relevant work. However, this metaphor preserves the crucial features of how the ANS is assumed to represent. Moreover, it provides some sense of why theorists have disagreed on what the representational capacities of the ANS might be. If we focus on the fact that the ANS discriminates cardinalities in a manner that exhibits magnitude and distance effects, then it is natural to suppose, along with Carey, that it represents *approximate* cardinalities. However, if one instead focuses on the *continuous* nature of the states of the accumulator, following Gelman and Gallistel, then one may be led to suppose that states of the ANS structurally resemble and represent real numbers. We will return to the question of whether the ANS plausibly represents numbers of any kind in Section 3.

Finally, we emphasize that the accumulator is widely assumed to conform to the same principles ensuring that count words in natural language represent cardinality. Quoting Carey (2009a, p. 132):

---

[7] Though see Carey (2009a) and Clarke and Beck (2021) for nice overviews.
[8] See Carey (2009a).

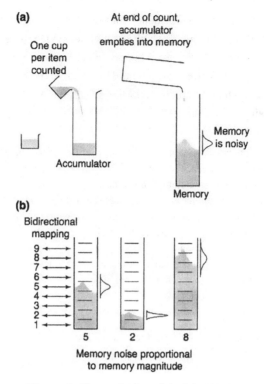

Figure 2 Theoretical model of the ANS.

The accumulator model instantiates the counting principles that ensure that [count] lists also encode number. In the accumulator representations, the successive states of the accumulator play the same role as successive number words in the list – as mental symbols that represent numerosity. States of the accumulator are stably ordered, gate opening is in correspondence with individuals 1–1 in the set, the final state of the accumulator represents the number of items in the set, there are no constraints on individuals that can be enumerated, and individuals can be enumerated in any order.

As we'll see, it is thanks to this supposed correspondence between ANS states and count words that, according to some Mainstream accounts, count words inherit their denotations from such states.

2.3.2 The Small Number System

In addition to the ANS, a second core system is widely assumed to be relevant to number cognition, what we'll call the "Small Number System" (SNS). However, there is considerable disagreement concerning how precisely to characterize it. Here are three prominent characterizations:

- *The Object Tracking System*: A system employed by the visual system for focusing attention on up to three or four objects at a time. It does so, thanks to the **object-indexing system** [Leslie et al. (1998); (Scholl and Leslie (1999)], which incorporates up to four symbols, or indexes, acting like pointers to the objects attended to, based on their spatial and temporal properties.
- *The Parallel Individuation System*: A domain-general capacity to construct and manipulate mental models of objects [Simon (1997); Le Corre and Carey (2007); Spelke et al. (2022)]. Mental models consist of distinct symbols, one for each object represented, held briefly in working memory. Consequently, only up to three or four objects are represented at any time.
- *The Subitizing Module*: An innate, domain-specific system responsible for precisely representing cardinality properties – oneness, twoness, threeness, and possibly fourness – without counting [Hurford (1987), Margolis and Laurence (2008); (Margolis (2020)]. There is a small stock of representations which are causally responsive to "numerical quantities" [Margolis and Laurence (2008); (Margolis (2020)]. For example, in the presence of a pair of shoes, this system immediately registers the twoness of the shoes.

On all these proposals, the hypothesized system *tracks* small cardinalities – its states causally covary with the number of items presented, up to three or four. However, there is significant disagreement regarding whether cardinality is *represented*. According to the first two proposals, the number of symbols or indices merely covaries with the number of objects detected. Only the last, subitizing module hypothesis, assumes that the SNS also *represents* numerical properties as such. Regardless, on any of these views, the SNS is implicated in the possession of at most the first three or four number concepts. Indeed, as we'll see, possessing number concepts beyond these appears to mark a significant developmental milestone, one which may require transcending the repertoire provided by the SNS *or* the ANS.

2.4 The Developmental Sequence

So far, we have sketched two core systems widely believed to be shared by both humans and many nonhumans, both tracking broadly numerical properties. However, since much Mainstream NCR is concerned with how children acquire basic *conceptual* capacities, there has been extensive attention to early manifestations of the sorts of behaviors that would provide evidence of number concept possession – especially, early counting behavior. As vanMarle (2018, p. 233) notes:

Because learning to count is the first step children take toward a symbolic understanding of number, researchers often focus on this milestone as a way to shed light on the process(es) underlying the development of the number concept.

In what follows, we briefly sketch the relevant notion of counting, the principles commonly taken to underlie knowledge of counting, and the developmental sequence associated with the acquisition of this ability.

Benacerraf (1965) distinguishes two forms of counting: **intransitive counting**, or reciting count words in their canonical order, typically starting with 'one' or its equivalent, and **transitive counting**, or determining the cardinality of a collection by establishing a one-to-one correspondence between its members and a sequence of count words. On one natural characterization, transitive counting is a procedure or routine for answering 'how many'-questions [Snyder et al. (2018a)]. Specifically

If asked "How many Fs are there?":

i. Isolate the Fs from the non-Fs.
ii. Establish a bijection between the Fs and an initial segment of the words in the count list $\langle \alpha_1 \ldots \alpha_n \rangle$ by reciting (possibly non-verbally) the words in order, starting with α_1 and correlating each F with a unique word in the list.
iii. If α_k is the final count word resulting from (ii), then answer 'There are α_k Fs'.

Suppose, for instance, one is asked 'How many Elmos are there?'. And suppose that, in response, one is able to establish a bijection between a set of four Elmos and English count words 'one', 'two', 'three', and 'four'. Then by the third step of the transitive count routine, one should answer the initial 'how many'-question with 'There are four Elmos'.

Given the presumed centrality of transitive counting (henceforth 'counting'), it is unsurprising that developmentalists seek to uncover the knowledge tacitly required to possess this ability. To that end, Gelman and Gallistel (1986; henceforth 'G&G') postulate five principles, known as **the counting principles**, three of which are purportedly necessary for counting correctly. Paraphrasing G&G, these are:

The One-to-One Principle: Each item to be counted should be counted once, and only once.
The Stable-Ordering Principle: When counting, count words should occur in a stable, and thus repeatable order.

The Cardinal Principle: The final count word used in counting "represents the number of items in the set," i.e. its cardinality.[9]

Intuitively, One-to-One guarantees the second step of the count routine, by matching each item being counted with a unique count word. Specifically, it prevents mapping, say, two Elmos to 'one', or one Elmo to 'three' but no Elmo to 'two'. In contrast, Stable-Ordering ensures that multiple performances of the counting routine on the same items will return the same answer. Specifically, it prevents the count list from occurring in different orders, whereby on one instance we intransitively count 'one', 'two', 'three', but on the next we count 'two', 'three', 'one'. Finally, the Cardinal Principle ensures that the answer given in the third step correctly answers the 'how many'-question posed in the first step.

The counting principles are often treated as a starting point for Mainstream accounts of number concepts. One reason is that they characterize what is presumed to be the first clear *manifestation* of those concepts. Since the task is linguistically mediated – it involves the use of number words – the comprehension and meaningful production of the relevant utterances would seem to provide good evidence for the possession of corresponding concepts (Section 2.2). However, we should be mindful not to assume that the first manifestation of this conceptual competence implies that counting is required for the acquisition of number concepts, since there are views – such as nativist proposals – on which children come to possess number concepts by other means. Nonetheless, as we'll see in Section 3, some very influential views maintain that learning to count is central to the *acquisition* of number concepts corresponding to the first few count words. Quoting vanMarle (2018, p. 132) again, according to these views, on learning

> how to implement the counting routine, ... [children] can determine the cardinal value of any set. This understanding, obviously, is very powerful. In acquiring it, children have done what it took centuries for philosophers and mathematicians to formalize: they have come to understand the system of positive integers (Carey, 2009).

This embodies a number of implicit philosophical assumptions, which we will revisit in Section 4. For now, however, we sketch what the developmental timetable – a widely accepted characterization of some of the central milestones involved in the early manifestation of numerical competences.

---

[9] Gelman and Gallistel (1986, p. 79).

The earliest parts of the timetable explicitly concern counting. Beginning at around age two, children learn to intransitively count. In doing so, they satisfy Stable-Ordering. At around this same time, they become competent in matching count words one-to-one with items in collections being counted, thus satisfying One-to-One. Thus, they appear to exhibit mastery of two of the counting principles very early in the developmental sequence. Interestingly, it is not until roughly two years later that children seemingly exhibit grasp of the Cardinal Principle.

In addition to studying how children learn to count, developmentalists have also paid extensive attention to the emergence of abilities to competently use count words in the course of solving various sorts of tasks. This aspect of the timetable, which occurs once children have learned to count (in)transitively, is standardly divided into **Knower-levels**. Though many tasks have been used to study this period in development, perhaps the most widely cited is Wynn's (1992) Give-a-Number task (**Give-N task**). Here, the child is asked by the experimenter to hand over a specific number of items – e.g. one cookie, two cookies, or five cookies. In the context of this task, children between approximately two and four years of age exhibit a stage-like progression in performance that has struck many NCRs as remarkable for being laborious, piecemeal, and slow. Carey (2009a, pp. 297–298) nicely describes the early phases of this progression:

> First, children are no numeral-knowers – they cannot even reliably give one object when asked for it. Between 24 and 30 months of age, most English-learning children become "one"-knowers. They can reliably give one object but hand over a random number of objects (always greater than one) when any other numeral in their count list is used in the request. They are in this stage for 6 to 9 months. They then become "two"-knowers (can reliably give one or two objects; chose a random larger number for any other numeral), and then "three"-knowers. Although it is much rarer, "four"-knowers have also been observed.

Collectively, 'one'-knowers, 'two'-knowers, 'three'-knowers, and 'four'-knowers are called **subset-knowers**, because they can reliably perform the Give-N task for a proper subset of their count list. Interestingly, there appear to be no *mere* subset-knowers for count words larger than 'four'. Rather, at around $3\frac{1}{2}$ or 4, children become capable of performing the Give-N task for *all* count words, up to the limit of their count list. For example, a child who can count up to ten will solve the Give-N task for requests up to and including 'ten'. Such children are called "cardinal principle knowers" (**CP-knowers**), because their behavior suggests tacit knowledge of the Cardinal Principle. As we will see in Section 3, however, there is considerable disagreement regarding how best to explain this transition, and in particular what tacit knowledge CP-knowers possess.

While much of Mainstream NCR has focused on numerical capacities manifested prior to the age of four, there is also research on the subsequent development of elementary arithmetic, such as solving basic equations like $3 + 2 = \_$. Although we focus far less on this research here, it is important for our purposes in Section 4 to appreciate that such developments are presumed to rely crucially on a prior capacity to *count*. For example, Butterworth (2005) describes the earliest development of arithmetic as typically happening at around age five or six in the United States, and as depending importantly on recognizing a link between (i) count words and their symbolic counterparts, e.g. '1', '2', '3', and (ii) arithmetic operations like '+' and counting operations, such as enumerating the result of joining two disjoint collections.

To summarize, here's a crude, approximate timetable for early childhood numerical development:

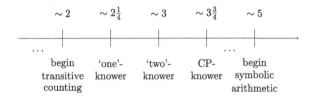

Clearly, counting plays a critical role in this developmental trajectory. Indeed, Butterworth (2005, p. 1) summarizes the development of elementary arithmetic this way: "Development can be seen in terms of an increasingly sophisticated understanding of [cardinality] and its implications, and in increasing skill in manipulating [cardinalities]." As we see it, the assumption that learning to count is a crucial precursor to acquiring other mathematical abilities is widespread within NCR. Indeed, continuing the quote from vanMarle (2018, p. 132) above: counting represents "the first symbolic mathematical knowledge that children acquire, and it is the foundation upon which all other formal math learning must be built." Little wonder, then, that counting occupies a central place within Mainstream NCR.

2.5 Some Explanatory Targets

So far we have articulated some of the central theoretical and empirical commitments underwriting Mainstream NCR. However, there is another, rather different sort of commitment that exerts a significant influence on how such research proceeds, namely: commitments regarding which phenomena are deemed explanatorily most central. Specifically, by our estimation, the the following three ontogenetic issues are central explanatory targets – as well as central foci for disagreement – within Mainstream NCR.

- *How do children come to possess low-value number concepts?* For example, how does an anglophone child acquire concepts expressed by the words 'one', 'two', ..., 'ten'? Assuming these concepts also correspond to those expressed by count words in other languages, the question can be formulated as: How do children acquire **low-value** count concepts? Moreover, on the apparently widespread assumption that such concepts represent natural numbers (or positive integers), the issue can be reformulated as: How do children acquire low-value natural number concepts – ONE, TWO, THREE, and so on?

- *How do children come to possess the concept NATURAL NUMBER?* In addition to acquiring concepts which, by hypothesis, denote specific natural numbers, there is a widespread interest in the issue of when, and how, children acquire a concept that denotes natural number, as such. Typically, it is assumed that this happens at a relatively late stage in conceptual development [Carey (2009a)], if indeed it happens at all [Rips et al. (2008)]. Moreover, it is almost invariably assumed to rely on the prior acquisition of concepts denoting specific natural numbers.

- *How do children acquire (tacit) representations of broadly numerical principles?* We have already mentioned Gelman and Gallistel's (1986) counting principles. Additionally, as we will see later, NCRs are also keen to understand when and how we acquire an understanding of arithmetic principles, such as those governing successor, as well as other principles of foundational significance, such as those governing equinumerosity. Such principles are often deemed central to NCR for at least two reasons. First, on certain widespread assumptions, it would be in virtue of grasping such principles that one could acquire the concept NATURAL NUMBER. Second, such principles are often taken to explain aspects of numerical behavior – e.g. children's coming to understand that there is no largest natural number (Section 3.5.3).

These three issues constitute the focus of much of what is to come. We start in Section 3, by considering some of the more influential attempts to explain the acquisition and manifestation of competence with number concepts.

3 The Acquisition Problem

When and how do children come to possess number concepts? Specifically, how do they come to possess concepts necessary for understanding low-value count words? In this section, we first develop a rough taxonomy of competing developmental hypotheses, and then discuss the main kinds of views in greater detail, along with challenges they confront.

3.1 A Taxonomy of Developmental Hypotheses

One common way of categorizing developmental hypotheses is by the innate cognitive resources they assume. With theories of number concept acquisition, those resources may be divided into two kinds: innate cognitive mechanisms, and innate number-specific representations. The options concerning both vary from relatively austere proposals assuming no number-specific innate resources to those assuming a rich innate endowment of number-specific cognitive resources.

Let's start with the most austere of extant views: those associated with **empiricism** within developmental psychology [Mix and Sandhofer (2007)]. Though they differ in details, such views exhibit a characteristic signature. First, they deny the existence of innate representations of number, numerical properties, or number-specific cognitive mechanisms. Instead, they typically assume just two main kinds of innate cognitive systems, namely:

- **Sensory-perceptual systems** underwriting vision, audition, olfaction, and other perceptual modalities.
- **Domain-general learning systems** operating similarly across a broad array of learning domains, including arithmetic, biological categories, theory of mind, the motion of physical objects, and so on.

Following Margolis and Laurence (2013), we call this **the Empiricist Base**. Within NCR, empiricists typically aim to explain the acquisition of number concepts without recourse to innate resources other than these.

One nonempiricist hypothesis, often associated with Chomsky, is that in addition to the sorts of mechanisms outlined above, acquiring number concepts further requires possessing an innate **language faculty**, and thus knowledge of the sorts of recursive morphosyntactic principles characteristic of transformational syntax [Chomsky (1987); Chomsky (2014)]. As we'll see, certain Mainstream NCRs entertain similar suggestions (Sections 3.4 and 3.5), and most assume an innate language faculty. However, advocates of the Mainstream frequently supplement the Empiricist Base with two further core systems specifically relevant to number cognition: the ANS and the SNS.

Against this backdrop, major differences between Mainstream approaches to number concept acquisition can be characterized by their responses to the following questions:

(Q1) Which number-relevant system(s) are centrally implicated in the acquisition of number concepts?

(Q2) What sorts of representations do these systems deploy?

(Q3) Are there additional innate number-specific cognitive systems that play
a role in the development of early number cognition?

Proposals answering (Q3) negatively may be divided into three mains sorts.

ANS-Dominant Models: In addition to language, perception, and domain-
general learning, acquiring number concepts relies crucially on the operations
of ANS.

SNS-Dominant Models: In addition to language, perception, and domain-
general learning, acquiring number concepts relies crucially on the operations
of SNS.

Hybrid Models: In addition to language, perception, and domain-general
learning, acquiring number concepts relies crucially on both the SNS *and*
ANS's capacities to represent number.

To be clear, most of the views considered below maintain that the various
systems mentioned play *some* role in number concept possession. Thus, the
distinctions proposed here are a matter of degree, not necessarily kind. Furthermore, within each view, there are significant disagreements regarding what
states of the relevant systems represent. As we'll see, this makes a significant
difference to the kinds of challenges such views confront.

Finally, those answering (Q3) affirmatively maintain that none of the aforementioned views are adequate to explain the acquisition of number concepts.
Rather, as we'll see (Section 3.3), they argue that a further system of mental
representations is needed, which provides resources for representing natural
numbers as such.

We turn now to the most prominent Mainstream views, starting with ANS-
dominant models.

3.2 ANS-Dominant Models

According to the kind of view defended by Dehaene (2011), Gallistel and
Gelman (1992, 2000), and Wynn (1992, 2018), number concepts originate in
numerical representations made available by the ANS. Thus, analog magnitude
representations form the primary cognitive foundation for all numerical knowledge. Specifically, the chief contention is that, in the course of development,
children grasp the meanings of number words – and, thus, come to possess corresponding concepts – as a result of mapping number words to ANS states with
the appropriate contents. For example, the child acquires the meaning of 'two'

by mapping that word to a state of ANS with the appropriate representational content, presumably concerning twoness. Thus, the following principle appears central to ANS-dominant models:

The Inheritance Thesis: Count words inherit their denotations from states of the ANS to which they are mapped.

According to this thesis, the count word 'two', for instance, denotes a content also denoted by a corresponding state of the ANS, in virtue of there being a mapping from the latter to the former.

To illustrate, consider one highly influential version of such a view, due to Randy Gallistel and Rochel Gelman [henceforth "G&G"; Gallistel et al. (2006)]. According to G&G, states of the ANS – specifically, of the accumulator – are presumed to be continuous and to represent continuous magnitudes analogically. Consequently, G&G assume that such states represent real numbers in much the same way that, say, the number line does.[10] For our purposes, the important question is how, on this assumption, such representational states can be used to learn the meanings of the count words, or concepts they express.

On G&G's account, since analog magnitudes already represent real numbers, the problem of learning count words consists in mapping those words to the positive integers embedded within the reals. Indeed, Gelman and Gallistel (1986) maintain that both the counting principles themselves and certain features of natural language allow children to solve this problem. Specifically, they claim the accumulator device in some sense "embodies" or "instantiates" the counting principles. Thus, G&G propose that by learning the verbal count routine, children also come to tacitly recognize a correspondence between count words and states of the ANS, thereby inferring that both represent the same things. Crucially, because count words are discrete, natural language acts as a kind of "filter" on which ANS states such words correspond to. Specifically, they will correspond to states representing the *positive integers* among the reals, as Gallistel et al. (2006, p. 19) explain:

> [T]he integers are picked out by language because they are the magnitudes that represent countable quantity. Countable quantity is the only kind of quantity that can readily be represented by a system founded on discrete symbols, as language is.

Thus, according to G&G, it is in virtue of the discrete character of count words that they inherit as their denotations not just any subset of the reals, but precisely those corresponding to the natural numbers.

---

[10] See Clarke and Beck (2021) for critical discussion of this assumption.

What should we make of this proposal? Broadly speaking, for ANS-dominant models to be adequate, two requirements must be satisfied:

The Representation Requirement: Some ANS states must have the same contents as count words, so that count words can inherit their contents from such states.

The Feasibility Requirement: It must be feasible for children to establish the salient mappings in the course of development.

The problem is that no extant ANS-dominant model appears to satisfy both requirements. Or so we'll argue.

As noted, various accounts of the ANS make quite different assumptions concerning the contents of ANS states. Indeed, it has been variously suggested that these represent:

- *real numbers*: Gallistel and Gelman (2000); Leslie et al. (2008)
- *rational numbers*, including the naturals: Clarke and Beck (2021)
- *approximate cardinalities*: Carey (2009a,b)
- *pure magnitudes*: Burge (2010).

To be clear, among these, only the ANS-dominant views of Gallistel and Gelman (2000) adopt the Inheritance Thesis. Regardless, our contention is that none of these views satisfy the Representation Requirement, and that some of them also fail the Feasibility Requirement. We begin with the former.

3.2.1 The Representation Requirement

According to the Representation Requirement, count words inherit their denotations from ANS states only if both denote the same things. On two of the major extant alternatives, these denotations are numbers, in the form of the reals or rationals. Thus, a significant problem with either suggestion is simply that numbers are not plausible denotations for count words. To see why, consider (3a,b).

(3)   a.   There are two Elmos on the table.

      b.   Two is an even number.

These different uses of 'two' not only plausibly belong to different morphosyntactic categories, they also clearly serve different semantic functions (Section 4.3). Specifically, whereas 'two' in (3a) is an adjective or determiner enumerating Elmos, 'two' in (3b) is a **numeral**, i.e. a name of a number, in this case two. Thus, of these two uses, only 'two' in (3a) deserves the moniker "count word," as only it is plausibly in the business of enumerating, semantically speaking. Furthermore, as we elaborate further in Section 4, contrastive pairs like (4a,b) strongly suggest that count words, unlike numerals or complex phrases like 'the number two', do not denote numbers.

(4) a. How many Elmos are on the table? {Two/??The (real/rational) number two}.

  b. Which number is Mary writing about? {Two/The (real/rational) number two}.

According to the Inheritance Thesis, count words inherit their denotations from ANS states. So, if the latter denote numbers, in the form of rationals or reals, then so should the former. In that case, the count word 'two' in (4a) would be coreferential with the numeral 'two' in (4b), and so we would expect (4a) and (4b) to be equally acceptable, contrary to fact. Consequently, views on which ANS states denote real or rational numbers are seemingly incompatible with the Inheritance Thesis.

Indeed, in this respect, views on which states of the ANS denote approximate cardinalities come closer to respecting the semantic evidence. On many extant analyses, 'two' in (4a) denotes the sort of thing answering 'how many'-questions, i.e. a *cardinality* [Moltmann (2013); Scontras (2014); Snyder (2017)]. Furthermore, it has been suggested, notably by Kennedy and Syrett (2022), that children assign these same entities as the denotations of count words at the earliest stages of numerical development. Thus, if ANS states denote cardinalities, albeit approximate ones, they would plausibly denote entities of the right sort, given the Inheritance Thesis.

But what *is* an approximate cardinality? Quoting Laurence and Margolis (2005, p. 221): "Instead of picking out 17 (and just 17), an Accumulator-based representation indeterminately represents a range of numbers in the general vicinity of 17." Fleshing this out somewhat, consider (5a–c):

(5) a. Exactly seventeen women attended the conference.

  b. Approximately seventeen women attended the conference.

  c. Seventeen women attended the conference.

Intuitively, (5a) is true if the number of women who attended the conference is *precisely* seventeen – no more, no less. In contrast, (5b) could be true if sixteen or eighteen women attended the conference, since these fall within a contextually given range which counts as close enough to seventeen. Now, within the context of performing the count routine, an utterance of (5c) is far more plausibly equivalent to (5a) than (5b). Indeed, if eighteen women actually attended the conference, then an utterance of (5c), like (5a), would be *false*, even though eighteen would be close enough to seventeen to count as approximately seventeen.

More precisely, suppose 'exactly' has something like the denotation in (6a), where '$[\![\alpha]\!]$' represents the denotation of 'α', 'n' ranges over natural numbers, 'S' ranges over sets, and '$|S|$' denotes the cardinality of S.

(6)   a. $[\![\text{exactly}]\!] = \lambda n.\lambda S.\ |S| = n$

       b. $[\![\text{exactly seventeen}]\!] = \lambda S.\ |S| = 17$

According to (6b), 'exactly seventeen' is true of those sets whose cardinality is strictly identical to 17. In contrast, suppose 'approximately' has the denotation in (7a), where '$\approx$' is a context-sensitive relation holding between numbers n and n' just in case n is among a range of numbers which are, in a given context, close enough to n'.

(7)   a. $[\![\text{approximately}]\!] = \lambda n.\lambda S.\ |S| \approx n$

       b. $[\![\text{approximately seventeen}]\!] = \lambda S.\ |S| \approx 17$

According to (7b), 'approximately seventeen' denotes sets whose cardinality is among a range of numbers which are, in a given context, close enough to 17, say $\{16, 17, 18\}$.[11]

Seen this way, views on which states of the ANS denote approximate cardinalities are incompatible with the Inheritance Thesis simply because count words do not plausibly denote *approximate* cardinalities. As (5a–c) demonstrate, there is little empirical plausibility to the claim that 'seventeen' is synonymous with 'approximately seventeen'. Rather, it is far more plausible that in the context of performing the count routine, 'seventeen' in (5c) denotes collections of *exactly* seventeen things [Horn (1972)]. In which case, 'seventeen' could not inherit its actual denotation from some ANS state, assuming such states denote approximate cardinalities.

Something similar can be said for views on which ANS states denote what Burge (2010, p. 482) calls **pure magnitudes**: "A pure magnitude is a magnitude not specific to any further type of magnitude such as spatial extent or size, temporal duration, weight, and so forth." That is, pure magnitudes are magnitudes divorced from any specific dimension of measurement, but which nevertheless permit basic arithmetic operations, such as addition, subtraction, multiplication, and division. They can also fall into ratios, in virtue of being subject to **the Eudoxus definition**. Let a and b be magnitudes of the same kind, e.g. two

---

[11]   As a reviewer observes, this does not account for the fact that the ANS treats values closer to the mean as more likely than those further away. For example, if there are approximately 100 people at the conference, where the salient range of values is $\{98,\ldots,102\}$, then by the lights of the ANS, it is more likely that there are 100 people than 98 or 102.

intervals of time, and let c and d be magnitudes of the same kind, e.g. two cardinalities. Then according to the Eudoxus definition, for any natural numbers n and m, $a : b :: c : d$ if and only if three facts obtain:

(ED1) $ma > nb$ if and only if $mc > nd$
(ED2) $ma = nb$ if and only if $mc = nd$
(ED3) $ma < nb$ if and only if $mc < nd$

Since pure magnitudes are divorced from specific dimensions of measurement, the Eudoxus definition also applies to them, only there is no distinction to be drawn between the kinds of magnitudes forming both sides of the proportion.

Burge (2010, p. 482) speculates that ANS states represent pure magnitudes:

> I believe that it would be fruitful to use Eudoxus' theory to understand representational contents of the perceptually based capacity to estimate numerosity. Such contents represent pure magnitudes as such. The magnitudes can fall into ratios, but they are represented in a way that is not specific as between continuous magnitudes and discrete magnitudes. At any rate, they do not represent discrete pure magnitudes – the numbers.

The important contention is that ANS states represent magnitudes which fail to distinguish between different dimensions of measurement, such as cardinality and length. Yet, as noted earlier, 'two' in (3a) most plausibly denotes a *cardinality*. Moreover, on standard accounts within linguistic semantics, what distinguishes cardinalities, as entities, from other kinds of magnitudes is precisely that the former encodes a particular, discrete dimension of measurement, namely cardinality [Scontras (2014); Snyder (2021a)]. As such, on the present proposal, count words could not denote *pure* magnitudes, and so could not inherit their denotations from ANS states.

3.2.2 The Feasibility Requirement

According to the Feasibility Requirement, count words inherit their denotations from ANS states only if there is a *feasible* mapping from count words to ANS states with an appropriate denotation. Here, ANS states are being individuated semantically, i.e. by their semantic contents. Thus, whether the Feasibility Requirement can be met will depend importantly on the presumed contents of ANS states. If, for instance, each state denotes a positive integer [Clarke and Beck (2021)], then a feasible mapping could presumably exist: just map each count word to a corresponding positive integer representation. However, if ANS states denote real numbers [Gallistel and Gelman (2000); Gallistel et al. (2006)], then it it is hard to see how such a mapping could exist, in principle.

The problem is evident: because there are infinitely many reals between any two naturals, the prospect of establishing a mapping between a given count word and its corresponding integer value will be infinitesimally small. Indeed, as Galistel and Gellman themselves later note:

> Learning the count words in a language, even in the presence of small sets of objects, will not help if the hypothesis space for possible (number) word meanings is the space of real values. Two real-valued measures of the same entity are, in general, infinitesimally likely to be exactly equal, and infinitesimally likely to have an integer value. Thus, the chance of a child entertaining an integer hypothesis would be infinitesimal. [Leslie et al. (2008, pp. 213–214)]

In light of this problem, why would G&G propose that ANS states denote real numbers in the first place? The answer, suggested by Gallistel and Gelman (2000), appears to be that this assumption is required to explain how the ANS represents duration as a continuous magnitude, as Clarke and Beck (2021, p. 12) explain:

> [G&G] reason as follows. First, because duration is a continuous magnitude, they infer that it cannot be represented by anything discrete. Therefore, the neural magnitude that represents duration must be continuous. Next, drawing on evidence in rats (Meck and Church, 1983), they infer that numbers are represented by neural "magnitudes indistinguishable from those which represent duration" (Gallistel and Gelman, 2000, p. 62). Thus, the ANS must also use a continuous neural magnitude. Finally, because real numbers are continuous, but integers are not, they conclude that the ANS must represent real numbers rather than just integers.

However, this line of reasoning is committed to a pair of related, problematic assumptions.[12] The first concerns the representational vehicles which may denote continuous magnitudes. Specifically:

Denotation-to-Vehicle: If a representational vehicle R denotes a value for some continuous magnitude, e.g. duration, then R is itself continuous.

This appears necessary to transition from the premise that duration – what's purportedly denoted by neural magnitudes – is continuous to the conclusion that the neural magnitudes representing duration – the vehicles themselves – are also continuous. However, the assumption is subject to clear counterexamples [Laurence and Margolis (2005)]. For example, digital clocks, digital thermometers, and symbols like 'π' and '$\sqrt{2}$' each involve discrete symbols

---

[12] For further critical discussion, see Laurence and Margolis (2005) and Clarke and Beck (2021).

representing continuous magnitudes – time, temperature, and irrational numbers, respectively.

The second problematic assumption is the converse of the first:

Vehicle-to-Denotation: If a representational vehicle R is a continuous magnitude, e.g. a spike of electrochemical activity, then what R denotes is continuous.

This appears to be required to transition from the empirical thesis that numbers (of some sort) are represented by continuous neural magnitudes, to the conclusion that the numbers so represented are reals. However, this assumption is also subject to obvious counterexamples [Laurence and Margolis (2005)]. For example, digital computers routinely use continuous magnitudes, such as voltage, to represent discrete values.

In summary, not only are there good reasons to deny that count words inherit their denotations from real-valued ANS representations, there were never especially good reasons to suppose that ANS states denote real numbers to begin with. We conclude that the Inheritance Thesis is implausible: given extant accounts of what ANS states denote, it is empirically implausible that count words inherit their denotations from such states.

3.3 Nativist Models

One influential response to the problems noted above is to posit an additional, innate system specifically dedicated to representing natural numbers. Perhaps the most well-known version of this view is due to Leslie, Gelman, and Gallistel [Leslie et al. (2008); Gelman et al. (2019)].

As Leslie et al. see it, the need to posit such a system is largely driven by empirical considerations regarding how children learn count words. Specifically, it is assumed that the number words children learn, and the concepts they express, denote natural numbers. Yet if ANS states represent something other than exact integer values – e.g. approximate cardinalities, or reals other than those corresponding to integers – then we are left without an explanation for the apparently obvious facts that (a) the number words children learn have exact integer values, and (b) the difference between the values denoted by adjacent count words is exactly one, as opposed to, say, approximately one. In that case, what sort of learning mechanism could explain how children learn count words?

In response, Leslie et al. (2008, p. 216) hypothesize an innate system, which they call a "minimal learning mechanism," purportedly satisfying four assumptions. Quoting Leslie et al:

(LM1) "There is at least one innately given symbol with an integer value, namely, ONE = 1."

(LM2) "There is an innately given recursive rule $S(x) = x + $ ONE. The rule S is also known as the successor function."

(LM3) "Each realized integer symbol is given a corresponding accumulator value. As with all such values, it is a noisy real. The difference between the accumulator values assigned to any two successive integer symbols is always approximately equal to the accumulator value assigned to ONE itself."

(LM4) "Inference mechanisms operating on these symbols support unbounded substitution. For any such symbol, N, N $*$ ONE = N $*$ ONE."

For our purposes, the important principles here are (LM1) and (LM2). Leslie et al. assume that word learning is a species of hypothesis formation and testing. Thus, if count words denote natural numbers, then learning the meaning of, say, 'two' requires possessing a system of symbols or representations in which suitable hypotheses can be formulated – e.g. that 'two' denotes the number two, or that 'two' denotes the successor of one. Moreover, on pain of circularity, such hypotheses cannot be formulated solely by using words in the child's natural language. For example, if one aims to form a hypothesis regarding the meaning of 'two', one must presumably *mention* – roughly, quote – the word. But the salient hypothesis cannot *use* 'two', since that would presuppose that the child has *already* learned the word in question [Fodor (1975)]. With this in mind, Leslie et al. propose (LM1) and (LM2) as a way of furnishing learners with the means of formulating hypotheses about the meanings of count words: symbols which are isomorphic to, and denote, the natural numbers.[13]

In our view, there are good reasons for doubting the plausibility of this proposal. One broadly philosophical objection concerns the assumption that count words denote numbers, which, as suggested above, is independently implausible. We return to this issue in Section 4.

Here, we focus instead on certain empirical challenges to Leslie et al.'s innateness hypothesis. Specifically, it seems to yield a variety of inaccurate predictions. Typically, innateness hypotheses are proposed to explain the manifestation of cognitive abilities which are:

(i) *Pan-cultural*: They are exhibited across (almost) all cultures.

(ii) *Precocious*: They manifest early in development, and are acquired with relative ease.

---

[13] (LM3) is intended to connect the innate integer system to the ANS, and thereby explain magnitude estimation judgments; and (LM4) is intended to permit arithmetic inference.

For example, both properties are much in evidence for visual perception and language. Absent any additional explanation, we would surely expect a similar pattern for counting, if all humans possessed an innate concept ONE, along with an innate representation of the successor function.

However, counting and count word use generally are neither pan-cultural nor precocious. As O'Shaughnessy et al. (2021) point out, there's considerable cultural variation in the manifestation of counting and count word use, ranging from societies like Canada or Scotland, where such capacities are endemic, to, e.g. the Pirahã culture, where it is largely or entirely absent. Moreover, even in industrial societies, learning the first few count words is a remarkably slow and piecemeal endeavor. As vanMarle (2018, p. 136) notes:

> If all children need to do is realize the correspondence between the verbal counting system and the nonverbal system, then it is unclear why it takes children so long to acquire the cardinal meanings of the count words and come to grasp the counting principles.

While we find this concern compelling, there is a familiar nativist response: the mapping problem just might be *difficult*, even if children possess innate representations of number.[14] Thus, Margolis (2020, p. 128) writes:

> [I]t doesn't follow that learning the meanings of the number words for small numbers is a trivial matter. There is still a very challenging mapping problem in which children have to determine that number words (and certain morphological features in language) pick out numerical quantities to begin with. And once they recognize that their interpretation should focus on numerical quantity, there is a further question about which numerical quantity is the right one for a given term. This is a daunting problem.

However, we think that this response – and, for that matter, standard formulations of the objection – fail to appreciate the core concern, namely the *contrast* between the rate at which count words and other words are learned during roughly the same period of development. The relevant ages – two to four years of age – fall within a period of linguistic development known as **the vocabulary explosion**, because children start learning words at a remarkably rapid rate. According to one estimate, during this period children learn approximately 8–10 words per day [McMurray (2007)]. Yet analogs of the challenges that Margolis notes plausibly exist *quite generally* for word learning, not just for count words. For example, when learning 'red' and 'blue' the child will need to work out which sort of property is denoted by each expression, namely *color*

---

[14] Indeed, in the previous section we discussed a nativist view that renders the mapping problem so hard as to render the learning of count words practically impossible.

properties, and also which specific properties among these are denoted. But if this is so, why should learning the first few count words, *in particular*, be so slow and piecemeal, if children already possess innate symbols representing natural number?

3.4 Hybrid Models

While ANS-dominant models seek to explain number concept acquisition primarily in terms of the ANS, hybrid models instead invoke both the ANS and the SNS. Though different versions exist, perhaps the most influential are due to Elizabeth Spelke [Spelke (2000, 2003, 2017); Spelke et al. (2022)]. Crucially, on her view, acquiring count concepts depends not only on the SNS and ANS, but also on a range of further systems, including the language faculty and the linguistic capacities it affords. Spelke's views have evolved considerably over the past two decades. In what follows, we focus exclusively on the most recent published version.

3.4.1 Spelke's Hybrid View

Spelke (2017, p. 148) summarizes her view as follows:

> I propose that natural number concepts arise through the productive combination of representations from a set of innate, ancient, and developmentally invariant cognitive systems: systems of core knowledge. In particular, natural number concepts depend on a system for representing sets and their approximate numerical magnitudes (hereafter, the Approximate Number System (ANS)) and a set of systems that collectively serve to represent objects as members of kinds. None of these core systems is unique to humans, but their productive combination depends on the acquisition and use of a natural language. Because both the core systems and the language faculty are universal across humans, and because children master their native language spontaneously, natural number concepts emerge universally, with no formal or informal instruction. Because language is unique to humans, so is our grasp of the natural numbers. Finally, because specific natural languages are learned, the system of natural number concepts is neither innate nor present in the youngest children.

This requires unpacking. Specifically, how does Spelke conceive of the ANS and the SNS, and what role is language assumed to play in furnishing children with count concepts? We'll consider these issues in turn.

According to Spelke, the ANS is a system for "representing sets and their approximate numerical magnitudes" – presumably their approximate cardinalities. In contrast, Spelke construes the SNS as a parallel individuation system, one which does not represent numerical properties, such as the number of entities in a visual array, but each individuated entity attended to – e.g. this cat and that dog – up to a limit of four objects. Each entity represented is associated with a distinct mental symbol [Hyde et al. (2017)]. More specifically, the system generates mental indices, or **object files**, corresponding to each entity represented.

As Spelke sees it, both the ANS and SNS are involved in the acquisition of count concepts. However, it is language that seems most central to her proposal. Specifically, according to Spelke (2017, p. 156), "children's discovery of the natural numbers depends on their mastery of the generative rules of their language," which they discover in four steps. The first three are intended to explain the child's slow and laborious mastery of the first three count words; the last is intended to explain how children expand their numerical vocabulary beyond 'three' (or its equivalent).

At the first step, children "construct a prolific and productive system of representations of object kinds, by mastering noun phrases that, in English, are composed of determiners and singular count nouns . . . (e.g. *a cup*, *the cat*, *your hand*)" [Spelke (2017, p. 157)]. The idea appears to be that children first master singular noun phrases (NPs), which specify entities of a given kind, by "linking each phrase to representations from three core systems: systems for representing cohesive, bounded objects, for representing animate agents and their actions, and for representing visual forms."

At the second step, "children learn natural language expressions that contain terms for individuals of different kinds (*the dog and the cat*) or individuals with different identities (*Evelyn and Simon*; *my cup and your cup*; *this dog, that dog, and the other one*)." Thus, by conjoining singular NPs previously mastered, children are able to form conjunctive NPs denoting pluralities of entities. These may be of the same kind, as with 'my cup and your cup', or different, as with 'the dog and the cat'. Importantly, Spelke assumes that conjunctive NPs are co-denoting with numerically modified NPs – e.g. 'my cup and your cup' is co-denoting with 'two cups', and 'this dog, that dog, and the other one' is co-denoting with 'three dogs'. Specifically, both kinds of phrases "refer to sets of two or three distinct individuals." In this way, Spelke proposes that children "learn part of the meanings of the first number words."

At the third step, children establish mappings from NPs involving the first three count words to two sorts of mental representations: (i) "representations of numerical magnitudes, delivered by the ANS," and (ii) "representations of arrays of 1–3 individual objects," delivered by the SNS. The mapping to ANS representations is intended to enable children to "discover that expressions like *three dogs* and *Rover, Fido, and Lassie* refer to sets of a larger numerical magnitude than the sets picked out by expressions like *two dogs* or *Rover and Fido.*" Thus, Spelke seemingly assumes that the ANS is capable of comparing the cardinal sizes of the sets denoted by such expressions, at least for those denoting sets of two or three objects. The explanatory role of the mapping from NPs to "representations of arrays of 1–3 individual objects" is less clear. It appears intended to accommodate the numerical *exactness* of NPs, such as 'three dogs' – a kind of exactness that is not readily explained by the ANSs approximate representations. Hence, by virtue of establishing this pair of mappings, Spelke proposes that:

> Children can appreciate that 'three dogs' refers to a set that is composed of exactly three distinct individuals and that is numerically larger than sets that can be referred to as 'two dogs'.

However, limits on the parallel individuation system and the ANS prevent similar comparisons for larger sets. Thus, more is required to explain how children expand their numerical vocabulary beyond 'three'.

The final step in Spelke's account proposes that children expand their numerical vocabulary by applying "the grammatical rules for forming expressions that refer to two or three individuals (*this dog and that one*; *two dogs*) so as to form expressions that refer to two or three sets of individuals (*three cows and one more, two geese and three ducks, two groups of three puppies*)." Specifically, Spelke suggests three ways in which children may "construct representations of larger numbers from representations of smaller ones." The first is through appending 'and one more' to a numerically modified NP, as with 'three cows and one more', thus forming a phrase apparently co-denoting with 'four cows'. The second is through conjoining numerically modified NPs, as with 'three cows and two horses', which, according to Spelke, "designates a set composed of a specific number of animals: *five animals*." Finally, we may use **group nouns**, such as 'group' or 'collection', to collect pluralities of things into individuated, and thus countable, entities [Landman (1989); Snyder and Shapiro (2022)]. To use Spelke's example, whereas we can apparently use 'three horses and three cows' to denote the same set denoted by 'six animals', we can alternatively use 'two groups of three animals' to apparently represent this same set.

This last example illustrates two ways of apparently referencing sets – by conjoining numerically modified NPs or by using group nouns. According to Spelke, these encode different arithmetic operations – addition and multiplication, respectively. In fact, Spelke intimates that in virtue of encoding such operations, speakers may come to recognize that the natural numbers are *infinite*. Quoting Spelke (2017, p. 158):

These new expressions designate sets whose numerical magnitudes are the sum or product of the magnitudes of the sets that compose them: From *three horses and two dogs* comes *three animals and two more*; from *three horses and three cows* comes *two groups of three animals*. By forming and interpreting these expressions, children gain the insight that numbers can be added and multiplied to form other numbers. Just as *three* can be applied to dogs, barks, and jumps, it can be applied to numbers: *three pairs, three dozen, three hundred*. And just as *and one more* can be added to a singular expression, it can be added to composed numerical expressions recursively, in principle without limit.... As Chomsky (1987) proposed, the productive rules for forming an infinite set of discrete, natural language expressions may underlie the discovery of the infinite series of natural numbers.

In summary, "the productive rules of natural languages, together with the resources of core knowledge, allow children to gain two critical insights": (i) numerical expressions "can be composed to express new numbers," and (ii) by forming complex, numerically modified NPs, "numbers can be added and multiplied to form other numbers."

3.4.2 Troubles for Spelke

In our view, the critical problem with Spelke's (2017) proposal is that it mischaracterizes the developmental target – count words that children come to understand. To succeed, Spelke's proposal should explain how children learn meanings that count words actually have. However, the meanings Spelke assigns to numerically modified NPs, such as 'two dogs', are simply not plausible, empirically speaking. So, while Spelke's account could perhaps explain how agents learn the meanings of expressions in *some* possible language, we maintain that it fails to explain how children learn numerical expressions in any extant natural language, such as English.

To appreciate the point, we need to say a bit more about Spelke's semantic assumptions. Though these are largely implicit in her presentation, she appears to suppose something like Chierchia's (1998a,b) semantics for nouns. This is illustrated in Figure 3, where '*a*', '*b*', and '*c*' denote **urelements**, and arrows represent the subset relation ($\subseteq$).

According to this analysis, whereas the singular noun 'dog' denotes urelements, in the form of dogs, the plural noun 'dogs' denotes sets resulting from taking the union of their unit sets.[15] The question here is how this analysis may be extended to provide a semantics similar to that plausibly assumed by Spelke.

---

[15] Technically, in order for the semantics described below to be compositional, following Chierchia (1998a), one needs to assume what is sometimes called "Quine's trick," identifying urelements with their unit sets.

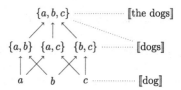

Figure 3 A set-based semantics for nouns.

First, consider definite singular NPs, such as 'Rover', 'the dog', or 'this dog'. Since these denote particular objects of a given kind, according to Spelke, such expressions are singular terms, referring directly to those entities. This is reflected in (8), where '*r*' represents Rover.

(8) ⟦Rover/the dog/this dog⟧ = r

Second, consider conjunctive NPs, such as 'Rover and Snoopy' or 'this dog and that dog'. These denote sets of entities referenced by each conjunct, according to Spelke, as suggested in (9).

(9) ⟦Rover and Snoopy⟧ = $\{r, s\}$

Hence, the most straightforward compositional analysis would have it that 'and' in the expressions of interest denotes set-union, as suggested in (10).

(10) ⟦and⟧ = $\lambda S.\lambda S'.\ S \cup S'$

Third, consider numerically modified NPs, such as 'two dogs'. These are co-denoting with conjunctive noun phrases, according to Spelke. Hence, 'two dogs' should share a denotation with, e.g. 'Rover and Snoopy'.

(11) ⟦two dogs⟧ = $\{r, s\}$

Fourth, consider elliptical phrases like 'one more (dog)', as in 'two dogs and one more (dog)'. Since the latter denotes a set consisting of three dogs, according to Spelke, the elliptical phrase presumably denotes a unit set of the same kind denoted by the preceding NP. Combining this with the previous denotations for 'and' and 'two dogs' thus results in (12), where '*t*' represents a third dog.

(12) ⟦two dogs and one more (dog)⟧ = $\{r, s, t\}$

Similarly, (10) and (11) jointly predict that the conjoined numerically modified NP 'three dogs and three cats' denotes a set containing three dogs and three cats – the same set denoted by 'six animals', according to Spelke.

(13) ⟦three dogs and three cats⟧ = ⟦six animals⟧ = $\{r, s, t, a, b, c\}$

In this respect, conjoined numerically modified NPs can perhaps be seen as encoding addition, as Spelke suggests. Finally, consider group NPs, such as 'group of animals'. Since 'two groups of three animals' is co-denoting with 'three dogs and three cats', according to Spelke, group NPs presumably denote sets of things described by the accompanying noun. For example, 'group of animals' presumably denotes a set of animals. This, combined with Spelke's assumption that conjunctive phrases of the form '*a* and *b*' are co-denoting with 'two *F*s' whenever *a* and *b* are both *F*s, may imply that 'two groups of three animals' denotes the union of two sets consisting of three animals, thus leading to the same denotation given in (13). In this way, numerically modified group NPs can perhaps be similarly understood as codifying multiplication, as Spelke suggests.

Though we think that the preceding analysis is the one most likely intended by Spelke, it should be evident that it cannot be correct as an account of the actual denotations of English expressions. The most significant problem concerns the assumed denotations of numerically modified NPs, such as 'two dogs'. Contra Spelke, these cannot generally be co-denoting with conjunctive NPs, such as 'Rover and Snoopy'. Consider (14), for instance, where each name refers to a different dog.

(14)   Snoopy and Rover are two dogs, and so are Toby and Ulyses.

Spelke's analysis wrongly predicts that (14) should be false in this context, since the two conjunctive NPs would refer to different sets.

What's gone wrong? Conceptually speaking, the problem with (11) appears to be that it involves what Geach (1972) calls **quantificatious thinking**: that all expressions occupying subject position serve the same kind of semantic function, namely *referring*, and that the referents of such expressions are all of the same kind, namely *sets*. Yet as Frege made clear long ago, there is little empirical plausibility to the claim that obviously quantificational expressions, such as 'nothing' and 'everything', function as *singular terms* [Kratzer and Heim (1998, Ch. 6)]. In terms of semantic types, the problem is that 'two dogs', when occurring in subject position, has the wrong type to function this way.

(15)   {Two/All/No} dogs are on the porch.

Within contemporary semantic theory, the phrases underlined in (15) are standardly assumed to have the type of **generalized quantifiers**, denoting *sets of sets* [Barwise and Cooper (1981)]. For example, 'two dogs' denotes the set of pairs of dogs. Thus, whereas one such pair may include, say, Rover and Snoopy, it may also include others, say, Toby and Ulyses.[16]

Though the above objections principally target Spelke's assumptions regarding the semantics of numerically modified NPs, they also ramify for her ontogenetic story. Most obviously, they undermine Steps 2 and 4 of the proposal. At Step 2, children are supposed to "learn part of the meanings of the first number words," by recognizing that numerically modified NPs, such as 'two dogs', are co-denoting with conjoined NPs, such as 'this dog and that dog'. But since such expressions are *not* co-denoting, a child who underwent Step 2 would, in fact, seriously *misunderstand* numerically modified NPs.

A similar problem afflicts Step 4. Here, children are presumed to expand their numerical vocabulary by applying "the grammatical rules for forming expressions that refer to two or three individuals." Importantly, this vocabulary expansion succeeds in virtue of these newly formed complex NPs combining denotations from previously understood, component NPs: "The sets to which these noun phrases refer inherit the properties of the sets that are the referents of noun phrases containing the first three number words." Clearly, then, no vocabulary expansion could take place unless these component expressions are already understood. Yet for each kind of complex phrase Spelke deems relevant to Step 4, the component NPs are numerically modified NPs whose meanings are supposed to be grasped at Step 2. Again, however, numerically modified NPs are not plausibly synonymous with conjunctive NPs. Thus, without some further, independent account of how understanding the latter could lead to understanding the former, we have little reason to think that Step 4 is achievable.

To summarize: though Spelke's proposal offers an intriguing account of how children come to understand count words, we maintain that it suffers substantial theoretical difficulties. In Section 4.2.3, we critically assess a more fundamental assumption that, while widely adopted within the Mainstream, is prominently illustrated by Spelke's proposal, namely whether the denotations of number words and number concepts are best construed in set-theoretic terms. For the moment, however, we turn our attention to SNS-dominant models of the sort most notably defended by Susan Carey.

3.5 SNS-Dominant Models

Whereas the previous accounts of count concept acquisition assign a pivotal role to the ANS, SNS-dominant models instead emphasize the role of the SNS.

---

[16] Group nouns raise an analogous issue. Specifically, given a set-based semantics, group nouns only plausibly denote sets of sets. See Landman (1989) and Snyder and Shapiro (2022).

Though views of this sort come in different forms, by far the most influential is due to Carey (2000, 2009a,b), and so we focus on her proposal here. Like Spelke, Carey assigns a central role to the language faculty. As we will see, however, her account of the specific role language plays is quite different from Spelke's. In what follows, we first explain Carey's proposal, and then discuss some of the central difficulties it encounters.

3.5.1 Conceptual Continuity and Quinean Bootstrapping

As a preliminary, it is worth locating Carey's proposal within the context of a broader debate that has not figured prominently in our discussion so far, but which is central to developmental psychology, at least since Piaget – issues regarding the possibility of **conceptual discontinuities**. A long-standing anxiety, both in philosophy and psychology, concerns how it is so much as possible for an individual to learn novel concepts – to learn concepts they did not possess at some prior time. Prima facie, we routinely perform this feat. For example, when learning science or mathematics, students seemingly learn such concepts as ELECTRON, QUARK, PHONEME or POLYNOMIAL.

Some nativists – most famously, in recent times, Jerry Fodor – have maintained that appearances are misleading – that such apparently novel concepts are not, in fact, learned, but rather innate [Fodor (1981)]. Yet few developmentalists are much inclined to accept this sort of "mad dog nativism" [Cowie (1998)]. Rather, a specific model of concept learning has tended to prevail: one in which we learn new concepts via processes which combine antecedently available representations. To illustrate, consider the widespread suggestion that concepts are acquired via a process of inductive extrapolation involving hypothesis formation and testing [Fodor (1981)]. On such a view, for example, acquiring BACHELOR would result from forming and then confirming a hypothesis of something like the following form,

The concept BACHELOR applies to x if and only if (and because) x is ...

where the right-hand side of the hypothesis specifies conditions under which BACHELOR is correctly applied. Though the details of this proposal needn't detain us, three related demands of this view should be appreciated:

The Correctness Requirement: Learning BACHELOR requires that one already possesses whatever concepts are needed to formulate a correct hypothesis regarding the application of BACHELOR. For if one did not, the hypothesis could not be formulated.

The Non-Circularity Requirement: On pain of explanatory circularity, the concept being acquired cannot also be used on the right-hand side of the hypothesis. For example, the hypothesis could not be:

The concept BACHELOR applies to x if and only if (and because) x is a BACHELOR.

For this would presuppose that the agent already possesses the concept whose acquisition we seek to explain.

The Basing Requirement: The previous two requirements imply that learning a concept C requires the prior possession of whatever other concepts are required to accurately formulate C's application conditions. Suppose, for example, that the relevant hypothesis is:

The concept BACHELOR applies to x if and only if (and because) x is UNMARRIED and x is MALE, and x is an ADULT, and x is HUMAN.

Then, the learner must antecedently possess the concepts UNMARRIED, MALE, ADULT, and HUMAN. That is, the learner must antecedently possess a relevant **conceptual base**.

There are two further implications of the aforementioned. First, if BACHE-LOR were learned in the envisaged manner, then there's a sense in which the conceptual system the learner possesses prior to acquiring BACHELOR would already provide the resources for having all those thoughts involving BACHE-LOR. After learning BACHELOR, the agent might have a new representation type, namely BACHELOR. But if the concepts on the right-hand side of the hypothesis capture the application conditions for BACHELOR, then a thought with the same content can be had by using these other concepts instead. This is sometimes framed in terms of preserving **conceptual continuity** between the conceptual systems possessed prior to and after this instance of learning. More precisely, we can formulate the relation of continuity between successive conceptual systems, CS1 and CS2, as follows:

(CC) CS1 and CS2 are *conceptually continuous* if and only if all the thoughts expressible in CS2 are expressible using only the concepts of CS1. That is, the expressive power of CS2 is no greater than that of CS1.

Moreover, if we further suppose that concepts are *only* learned via the aforementioned method, then we appear committed to the following theses:

Primitive Basing: There is a base of primitive conceptual representations – representations that do not have other representations as proper parts.
Continuity Thesis: All acquired conceptual systems are conceptually continuous with the initial, primitive base system.

A commitment to Primitive Basing arises because if there were no such base, concept learning could not get off the ground. Indeed, every instance of concept learning would presuppose some prior instance, ad infinitum. Moreover, a commitment to the Continuity Thesis arises because, as indicated, the sort of concept learning envisaged preserves continuity at each step. Thus, chaining the above together, if all concepts are learned via the sort of constructive process outlined, then all possible human thoughts would be expressible in a base of primitive, unlearned concepts.

Though such a view has been defended by some very influential figures [e.g. Macnamara (1986)], it has striking implications that many have found implausible. With regard to mathematics, specifically, Jean Piaget famously observes:[17]

> [T]his would mean that a baby at birth would already possess virtually everything that Galois, Cantor, Hilbert, Bourbaki or MacLane have since been able to realize. [Piaget and Chomsky (1980, p. 26)]

In light of this, many developmentalists have denied the Continuity Thesis, seeking instead to provide accounts of conceptual change that allow for **discontinuities** – changes where some of the thoughts expressible in CS2 are not expressible using only the concepts of CS1. Carey herself seems to be developing a view of this sort by characterizing a kind of learning she calls **Quinean bootstrapping** – "Quinean" because she claims to detect the general idea in the work of W.V.O. Quine, "bootstrapping" because, as with the well-worn metaphor of pulling oneself up by one's bootstraps, the proposed process enables progress in a manner which, at first glance, appears impossible. As Carey (2000, p. 20) puts it:

> The metaphor captures what is hard about the process of creating new representational resources that are not entirely grounded in antecedent representations.

Specifically, according to Carey, bootstrapping permits the acquisition of a more powerful conceptual system on the basis of less powerful ones.

So what is Quinean bootstrapping, exactly? Though Carey's characterization is impressionistic, the following presumed features are clearly central. First, it is intended to effect conceptual discontinuities. Second, it is supposed to be a species of *learning process*. In particular, it is a computational process which involves operations on mental representations and preserves rational relations

---

[17] See Carey (2009a, p. 18) for further discussion.

between inputs and outputs. Moreover, it recruits a wide array of other cognitive processes combined in various ways, including analogical mapping, deduction, and hypothesis formation and testing.

Third, Quinean Bootstrapping is assumed to rely on the initial availability of what Carey calls a **placeholder system**: a system of explicit symbols, such as a list of words in a natural language, which initially is "at most only partially interpreted with respect to antecedent concepts." More specifically, the meanings initially attributed to the symbols in the placeholder system, by the learner, are exhausted by their (presumably nonsemantic) relations to each other, as opposed to being "interpreted in terms of already existing representations" [Carey (2009a, p. 246)].

Finally, according to Carey, Quinean bootstrapping is a kind of "modeling process," in that as a result of deploying the sorts of processes mentioned earlier, the learner is ultimately able to assign an interpretation to symbols in the placeholder system. According to Carey, it is by assigning such interpretations that the learner comes to possess new concepts.

3.5.2 Bootstrapping the Count List

Broadly speaking, on Carey's account, children acquire number concepts in the course of learning how to count. Her proposal may be divided into three stages, each corresponding to periods in the developmental timetable outlined in Section 2.4.

Stage 1, occurring around two years of age, consists in the child learning the initial placeholder system, i.e. the *count routine*. During this period, the child learns to count intransitively and to satisfy the One-to-One Principle. According to Carey, children do not use count words to express any specific concepts at this point. Rather, count words are merely uninterpreted, perhaps phonologically individuated, tokens. To the extent that they have any meanings for the child, these are initially exhausted by their relations to each other. Specifically, words in the count list "get initial interpretations as part of a placeholder structure, the count list itself, in which meaning is exhausted by the fact that the list is ordered" [Carey (2009a, p. 19)].

Stage 2, corresponding to the subset-knower stage, consists in acquiring an understanding of the meanings of the first four count words. According to Carey, this results from mapping each word to a "long-term memory model" – a representation created by the parallel individuation system and stored in long term memory. Carey speculates that such models represent sets while also containing symbols representing individuals – "{i}, {j k}, {m n o}, {w x y z}" [Carey (2009a, p. 19)]. Moreover, she proposes that the first four count words have their meanings in virtue of a computational procedure for assigning count words to sets:

> The child makes a working-memory model of a particular set he or she wants to quantify, for example {cookie cookie}. He/she then searches the models in long-term memory to find that which can be put in 1–1 correspondence with this working-memory model, retrieving the [count word] that has been mapped to that model.

Thus, in the mouths of subset-knowers, the meanings of the first four number words are determined by a computational procedure in which they participate: a procedure ensuring that each word can be applied to any set in 1-1 correspondence with an appropriate set-theoretic representation stored in long-term memory.

Stage 3, which centers on the transition to becoming a CP-knower, involves mastery of the Cardinal Principle (Section 2.4). To appreciate Carey's account, it is important to distinguish this from an initially similar principle, namely what [Fuson (2012)] calls "The Last Word Rule"

LWR: The last count word used in the counting procedure answers the relevant 'how many'-question posed when performing that procedure.

There is a crucial difference between these two principles. Whereas grasping the Cardinal Principle implies that the child understands the *meanings* of count words, grasping LWR does not. Rather, it merely implies an ability that might be, as it were, brute mechanical – the ability to respond with the appropriate word, in the appropriate circumstances. However, on empirical grounds, presented in Sarnecka and Carey (2008), Carey supposes that subset-knowers already understand LWR, and hence that grasping it cannot explain the transition to becoming a CP-knower.

We find an analogy to Searle's (1982) **Chinese room** helpful here. In this familiar thought experiment, a non-Chinese speaker is handed cards containing Chinese characters through a slot in a door. There are instructions in the person's native language for processing each card, producing further cards with Chinese characters, and passing these new cards through another slot. We may imagine that the person becomes so good at following these instructions that they process cards in a time comparable to a native Chinese speaker, even without the assistance of instructions. In that case, the two speakers might appear behaviorally equivalent: for each input, the two speakers return the same output, in a comparable time. Yet unlike the Chinese speaker, the non-Chinese speaker does not *understand* what is on the cards. Rather, they are merely following a procedure, albeit very well.

The analogy is this: In the context of counting, mastering LWR merely enables the child to deliver responses that are appropriate. Indeed, a child grasping LWR could be very good at performing the transitive counting routine without grasping the meanings of count words or possessing count concepts. In contrast, according to Carey (2009a), CP-knowers are like Searle's Chinese speakers – when grasping the Cardinal Principle, they don't merely respond appropriately, they also *understand* the meanings of the count words, and so possess count concepts. In making this claim, Carey appears to be applying similar considerations to those militating in favor of attributing number concepts to subset-knowers. For instance, the standard reason for attributing the concepts ONE, TWO, and THREE to three-knowers is that they manifest successful performance on Give-N and related tasks, up to the word 'three'. Analogously, since CP-knowers successfully perform such tasks for *all* count words on their count list, similar considerations suggest the possession of concepts for all those count words.

Suppose with Carey (2009a) that on becoming CP-knowers, children tacitly recognize that count words denote what Carey calls **cardinal values**, or collections of equinumerous sets. The challenge is to explain how they come to grasp the meanings of *all* words in their count list, and not just the first four. To this end, Carey invokes bootstrapping. Specifically, she proposes that children recognize a structural analogy between count words and the sets they denote. Count lists are **linearly ordered**.[18] That is, just as ordering the natural numbers by < results in the sequence

$$1 < 2 < 3 < \ldots$$

ordering count words by the relation of coming next in the count list, represented here as '$\prec$', results in a similar sequence:

$$\text{'one'} \prec \text{'two'} \prec \text{'three'} \prec \ldots$$

Now, consider the following relation on sets, which we call **cardinal successor**, represented as '$\ll$'.

(16)	$S \ll S'$ if and only if for some $x \notin S$, there's a one-to-one correspondence between $S \cup \{x\}$ and S'

According to (16), S' is the cardinal successor of S just in case adding a new object to S results in a set which is in one-to-one correspondence with S'.[19]

---

[18] Generally, linear orderings are transitive (if $x \leq y$ and $y \leq z$, then $x \leq z$), antisymmetric (if $x \leq y$ and $y \leq x$, then $x = y$), and total (either $x \leq y$ or $y \leq x$).

[19] A similar definition is suggested by Frege (1884).

Clearly, ordering sets by cardinal successor results in a linear ordering. For example, we have

$$\{a\} \ll \{a,b\} \ll \{a,b,c\} \ll \ldots$$

This appears to be the presumed structural analogy CP-knowers tacitly recognize: count words ordered by $<$ and the sets they are assumed to denote, ordered by $\ll$, form isomorphic structures [Cheung et al. (2017)].

Importantly, recognizing this analogy licenses what Carey (2009a, p. 289) calls "the crucial induction": "For any symbol in the numeral list that represents cardinal value n, the next symbol on the list represents cardinal value $n+1$." More precisely, Carey's comments suggest the following **metasemantic principle** governing the presumed denotations of count words.

(MP) For every count word α, there's a unique next count word β such for any set S in the extension of α, if an object x is not in S, then adding x to S results in a set S' in the extension of β.

In plain English, (MP) implies the conjunction of two claims:

The Morphosyntactic Condition: Every count word is followed by a unique next count word.

The Semantic Condition: If a count word is true of sets containing n-many things, the next count word is true of sets that result from adding an extra object.

Some clarifications are in order here. First, notice that the Morphosyntactic Condition implies that the count list is, in fact, *infinite*. Second, the Semantic Condition guarantees that all count words in the count list are provided an appropriate denotation, in the form of cardinal values. For example, given that 'four' denotes the collection of four-membered sets, the Semantic Condition guarantees that 'five' denotes sets resulting from adding a new member to those sets in the extension of 'four'. Third, as we discuss further in Section 4, this implies that count words denote *natural numbers*, if we further suppose the Frege–Russell characterization of the naturals. Fourth, the joint effect of the aforementioned three assumptions is that there are infinitely many count words, each denoting a natural number, and each such that the number denoted by the count word immediately following it is its unique cardinal successor. Put differently, the aforementioned assumptions jointly entail:

The Successor Principle: For each natural number, there is a unique natural number immediately following it.

And this, of course, is precisely what the Dedekind–Peano axioms characterizing successor entail. Thus, in combination with the aforementioned assumptions, (MP) implies that natural numbers conform to those axioms.

Under the aforementioned assumptions, then, if children become CP-Knowers as a result of learning (MP), then they should also come to tacitly know, or at any rate be in a position to infer, the Successor Principle. Accordingly, Carey (2009b, p. 251) describes the key cognitive difference between subset-knowers and CP-knowers as follows:

> [T]he difference between subset-knowers and cardinal principle-knowers is a (tacit) appreciation of how counting implements the successor function, precisely the induction posited in the above bootstrapping proposal.

The suggestion appears to be that in virtue of bootstrapping (MP), CP-knowers tacitly possess knowledge of successor, a powerful consequence of which is an ability to represent *every* natural number, in principle.

> When the child, at around the age $3\frac{1}{2}$, has mastered how the count sequence represents number, he or she can in principle precisely represent any positive integer. Carey (2009a, p. 328)

A similar idea is echoed in vanMarle (2018, p. 132):

> With this knowledge [of the Cardinal Principle], children presumably understand ... how to implement the counting routine, and in doing so, they can determine the cardinal value of any set.

Again, the presumption is that an ability to represent the full system of natural numbers, at least in principle, is concomitant with becoming a CP-knower, something which only makes sense if bootstrapping (MP) implies tacit knowledge of the Successor Principle.

Carey (2009a, p. 328) summarizes her account as follows: "I have argued here that the numeral list representation of number is a representational resource with power that transcends any single representational system available to prelinguistic infants." In other words, bootstrapping (MP) allows CP-knowers to have thoughts which are in principle inaccessible to pre-CP-knowers, namely those involving count concepts beyond FOUR. As a result, tacitly grasping (MP) constitutes a kind of conceptual *discontinuity*.

3.5.3 Troubles for Carey

Carey's rich, provocative account of number concept acquisition has been the focus of much discussion. Though, in Section 4, we critically assess some of its more general assumptions, e.g. that count words denote sets, and that natural numbers are collections of sets, in this section, we focus on a pair of issues. The first concerns the extent to which early CP-knowers understand the Successor Principle. The second concerns whether Carey's account delivers a conceptual discontinuity in the course of bootstrapping (MP).

Do Early CP-Knowers Tacitly Know the Successor Principle? As we have seen, Carey's bootstrapping seemingly predicts that becoming a CP-knower is a major milestone in mathematical development, in that it implies tacit knowledge of the Successor Principle. Cheung et al. (2017, p. 23) nicely describe this key assumption.

> According to some accounts [e.g. Carey (2009a)], the transition to becoming a CP-knower marks a profound inductive leap ... Critically, according to this account, this knowledge, which "turns a subset knower into a cardinal-principle knower" is thought to reflect implicit knowledge of the successor function (p. 673), as described by the Peano–Dedekind axioms.

If so, then one might reasonably test this prediction by attempting to determine whether early CP-knowers actually exhibit knowledge of successor. Thus, in a series of papers, David Barner and colleagues have argued on empirical grounds that no such knowledge exists, and hence, that there are good reasons to reject Carey's analysis.

By way of illustration, consider Cheung et al.'s (2017) Largest Number and Successor probes. In each, CP-knowers are asked open-ended questions relating to the infinity of the natural numbers. An example of a Largest Number probe would be "If I keep counting, will I ever get to the end of numbers, or do numbers go on forever?" and an example of a Successor Probe would be "If we thought of a really big number, could we always add to it and make it even bigger, or is there a number so big we couldn't add any more?" What they found is that early CP-knowers tended to answer such questions incorrectly. For example, they often say that there *is* a biggest number, or that numbers do *not* go on forever. Cheung et al. (2017, p. 32) summarize their findings as follows:

> This study is the first to document the complete developmental trajectory of successor function acquisition, and finds that it is highly protracted. Using a stronger test of successor knowledge, we found that not until after age 6 can children both identify the values of successors within their count list and reason about the successor function for all possible numbers.

That is, children reliably exhibit an understanding of the Successor Principle only approximately two years *after* becoming CP-knowers, strongly suggesting that CP-knowers do not grasp the Successor Principle in virtue of bootstrapping (MP). In fact, Carey herself concedes as much in a paper recently co-authored with Barner [Carey and Barner (2019, p. 813)]:

> [L]earning a verbal encoding of the successor function builds directly upon counting abilities found only in CP-knowers, not just on associations between words and magnitudes. Only by around age $5\frac{1}{2}$ years – 1 or 2 years after they become CP-knowers – do children exhibit evidence of having learned a fully recursive successor function that generates an infinite set of numbers.

So it would appear that it cannot be in virtue of inferring (MP) that children tacitly grasp the Successor Principle, by Carey's own lights.

The Cheung et al. results may initially appear devastating to Carey's proposal. However, we think that there are two reasonable grounds for resistance. The first centers on the notion of tacit (or implicit) knowledge assumed. Both Cheung et al. and Carey frame the issue as concerning whether children have "implicit knowledge of the successor function." Notoriously, however, the notion of implicit or tacit knowledge has been construed in importantly different ways; and while there is a familiar sense in which Carey's proposal implies tacit knowledge of the Successor Principle, it is not plausibly the sense being tested by the Cheung et al. studies.

As mentioned in Section 2, cognitive scientists often assume the existence of representational states which are tacit in the following sense:

Tacit Knowledge as Occurrent: A proposition p is tacitly known – or tacitly represented – by an agent A if there is a cognitive state of A that a) is individuated by the content p; b) can participate in causal interactions; and c) cannot be verbally reported.

Such "subpersonal" or "unconscious" states are posited in different regions of cognitive science, e.g. the study of visual perception and language, and figure centrally in causal explanations. However, it is far from clear that Carey is committed to CP-knowers having tacit knowledge of the Successor Principle in *this* sense. On her view, recall, what explains the target behaviors – e.g. the ability to succeed on the Give-N task for count words greater than 'four' – is a representational state encoding (MP). That is, it is this metasemantic principle that, on her view, is plausibly tacit in the aforementioned sense. In contrast, it would seem that the Successor Principle is tacit – or implicit – in some different and rather weaker sense. Specifically, it would appear to be tacitly known in the sense that given a state encoding (MP) – and given suitable background beliefs, and inferential capacities – the Successor Principle is *inferable* from (MP). Generalizing:

Tacit Knowledge as Dispositional: A proposition *p* is tacitly known – or tacitly represented – by an agent *A*, if, given *A*'s inferential capacities, *p* is inferable from their other representational states.

To illustrate, although Smith has no representational state representing that dogs do not have 20,000 legs, this is readily *inferable* from something that she does represent, namely that dogs have no more than four legs. Although more complex, the status of the Successor Principle in Carey's theory is plausibly analogous.[20]

We can now explain why the Cheung et al. study fails to test the relevant claim. As made clear above, on Carey's account, CP-knowers only tacitly know Successor in the *dispositional* sense – it is inferable from (MP) and other background assumptions. Yet the evidence given by Cheung et al. more clearly tests for tacit knowledge in the *occurrent* sense. Accordingly, Carey could have an explanation of the evidence brought forth by Cheung et al: early CP-knowers are simply yet to draw the required inference.

The second response to the Cheung et al study concerns the details of how one formulates (MP). Specifically, Carey (2009a) provides two different characterizations:

> For any symbol in the numeral list that represents cardinal value *n*, the next symbol on the list represents cardinal value *n* + 1. [Carey (2009a, p. 289)]

> [I]f "x" is followed by "y" in the counting sequence, adding a individual to a set with cardinal value x results in a set with cardinal value y. [Carey (2009a, p. 327)]

Despite their similarity, there is a significant difference between these formulations: whereas the former implies the existence of infinitely many count words and corresponding cardinal values,[21] the latter, as a conditional statement, does not. In fact, this second formulation is consistent with there being *no* count words, or cardinal values, at all.

More formally, we can represent Carey's two formulations as:

---

[20] Of course, these are not the only senses in which knowledge might be said to be "tacit." For extended discussion, see Ramsey (2022).

[21] Provided the familiar assumption that utterances of 'the *F*' either entail or presuppose the existence of *F*s [Chierchia (1998b)].

(IMP) For every count word α, there is another count word β immediately following α such that if α is true of a set S and $x \notin S$, then β is true of $S \cup \{x\}$.

(FMP) For every count word α, if there is another count word β immediately following α, then if α is true of a set S and $x \notin S$, then β is true of $S \cup \{x\}$.

Clearly, only (IMP) implies the existence of infinitely count words and corresponding cardinal values, in the form of sets. As such, only (IMP) is a viable candidate for codifying the successor function, as characterized by the Dedkekind–Peano axioms. After all, the latter jointly require the existence of a (unique) successor for *every* natural number, not just those which may be the denotations of a finite list of count words.

With this distinction in hand, a second reason Cheung et al.'s experiments might fail to undermine Carey's proposal is this: If (FMP) expresses the salient metasemantic principle, then there is clearly no reason to suppose that by learning that principle, children would thereby acquire tacit knowledge of the Successor Principle, even in the dispositional sense. Thus, a version of Carey's account formulated in terms of (FMP) would in no way be threatened by the Cheung et al results.

On the other hand, such an account would also clearly fail to establish that CP-knowers have tacit knowledge of *the Successor Principle*. Thus, Carey's account of how CP-knowers "can in principle precisely represent any positive integer" plausibly succeeds only if children actually infer the *inifinistic* metasemantic principle codified in (IMP). However, this proposal faces substantial challenges.

First, independent of and prior to the empirical findings of Cheung et al. (2017), there is no evident reason for thinking that children should infer (IMP), rather than (FMP), given the child's limited experience with counting. After all, both metasemantic principles predict the same results up to the limit of the child's count list. Specifically, both imply that words within the count list and their cardinal values are similarly ordered. Consequently, if one principle can explain the transition from being a subset-knower to becoming a CP-knower, then so can the other.

Furthermore, there are compelling, independent reasons for doubting the veracity of (IMP). Specifically, contrary to what Carey (2009a, Ch. 9) suggests, natural language count lists are not, in fact, infinite.[22] Because number word

---

[22] To quote Carey (2009a, pp. 335–336): "Over the next few years, the child extends the verbal numeral list to 'one hundred' and then 'one thousand,' beginning to command the base system, and coming to the realization that there is no highest numeral."

morphosyntax is not *recursive*, there are clear upper-limits on what number words can be generated within any natural language [Grinstead et al. (1997)]. Consequently, the characteristic feature of (IMP) – the Morphosyntactic Condition – is simply false.

> *The Morphosyntactic Condition*: Every count word is followed by a unique next count word.

Put in terms of a distinction familiar to philosophers of mathematics, given the limitations on count word morphosyntax, count lists are at most **potentially infinite**, as opposed to **actually infinite** [Schlimm (2018)]. In other words, while it may be that for every count word α, there *could be* a next count word β, it is patently not the case that for every count word α, *there is* a next count word β. Yet the Dedekind-Peano axioms require that natural numbers are *actually* infinite, not merely potentially infinite.[23]

Finally, even if children did infer (IMP) by recognizing a structural similarity between natural language count words and their purported denotations, this would imply knowledge of the Successor Principle only if natural numbers are identified with certain sets. As characterized by the Dedekind-Peano axioms, successor is a relation holding between natural *numbers*. Thus, (IMP) codifies that relation only if the Frege–Russell characterization of the naturals is correct, an assumption we will criticize at length, on independent grounds, in Section 4.2.3.

Bootstrapping and Discontinuity   Our second objection targets the programmatic aspirations of Carey's account. Recall that major motivation for bootstrapping number concepts is that such learning requires a conceptual discontinuity. In what follows, however, we argue for three claims:

i. The notion of a conceptual discontinuity, as Carey uses it, is equivocal between two readings: a cumulative reading and a distributive reading.

ii. Even on the assumption that she accurately characterizes both what the child learns, and the core systems involved in this learning process, there are good reasons to suppose that her account fails to deliver conceptual discontinuities in either sense.

iii. Given (i) and (ii), there is no reason to suppose that the transition to CP-knowing involves a conceptual discontinuity.

Two notions of conceptual discontinuity.   Earlier we characterized the notion of conceptual continuity as a relation between conceptual systems:

---

[23] For a nice explication of the distinction between potential and actual infinity, along with its relevance to number theory, see Linnebo and Shapiro (2017).

(CC) CS1 and CS2 are *conceptually continuous* if and only if all the thoughts expressible in CS2 are expressible using only the concepts of CS1. That is, the expressive power of CS2 is no greater than that of CS1.

Conversely, conceptual discontinuity between two systems is the failure of continuity – i.e. where the expressive power of CS2 is greater than that of CS1. Although such relations may obtain **synchronically** – e.g. between mine and your conceptual systems at a time – **diachronic** relations of conceptual continuity and discontinuity are of special interest to developmentalists, since they concern ways in which conceptual systems may evolve over time. It is important to see, however, that we can individuate conceptual systems in two quite different ways that lead to importantly different sorts of diachronic (dis)continuities.

According to the first, *cumulative* conception, a conceptual system is an agent's *total* conceptual resources at a time, considered jointly. Crudely, CS1 consists of *all* conceptual structures the agent possesses at $time_1$, while CS2 consists of *all* conceptual structures possessed at $time_2$. Thus, to say that CS2 is discontinuous with CS1 implies that the *total*, collective conceptual resources of the agent at $time_1$ are insufficiently powerful to permit thoughts the agent could have at $time_2$. Plausibly, it is this notion of discontinuity figuring most prominently in, say, the disputes between Piaget and Chomsky. For example, when Jerry Fodor asserts the 'impossibility of acquiring "more powerful" structures', he is denying this sort of discontinuity. Call this **the cumulative conception of discontinuity**.

However, on an alternative conception, rather than considering CS1 as a single system consisting of all conceptual structures jointly possesses at $time_1$, we instead think of CS1 as partitioned into multiple, *distinct* conceptual or representational systems – parallel individuation, approximate magnitude, theory of mind, language, and so on – at a time. To assert the existence of a conceptual discontinuity then implies that *each* of the agent's representational systems at $time_1$ is such that *it* is insufficiently powerful to permit thoughts that the agent could have at $time_2$. For example, the parallel individuation system might be insufficiently powerful to represent natural number, as might the ANS. Call this the **distributive conception of discontinuity**.

Below, we assess Carey's arguments for such a discontinuity in the case of number. For the present, however, we clarify the logical relations between these two notions of discontinuity. First, the distributive notion is *logically weaker* than the cumulative one: a cumulative discontinuity entails a distributive discontinuity, but not vice versa. That is, if all representational systems collectively comprising CS1 cannot *jointly* express what CS2 can, then none of those systems *individually* can, but not necessarily vice versa. Conversely, the *absence* of a distributive discontinuity implies the absence of a cumulative discontinuity as well. Specifically, CS1 will fail to be distributively discontinuous with CS2 only if at least one of the individual systems that comprise CS1 can express what CS2 can. Yet if a proper part of CS1 can express what CS2 can, then so can the whole CS1, of course.

Which sort of conceptual discontinuity does Carey's account secure? The answer, we suggest, is "Neither." First, as evidenced from the following passage, Carey (2011, p. 118) appears only to be arguing for a *distributive* discontinuity with respect to the representation of exact cardinality:

> CS2 is qualitatively different from each of the CS1s because none of the CS1s has the capacity to represent the integers. Parallel individuation includes no symbols for number at all, and has an upper limit of 3 or 4 on the size of sets it represents. The set-based quantificational machinery of natural language includes symbols for quantity (plural, some, all), and importantly contains a symbol with content that overlaps considerably with that of the verbal numeral "one" (namely, the singular determiner, "a"), but the singular determiner is not embedded within a system of arithmetical computations. Also, natural language set-based quantification has an upper limit on the sets' sizes that are quantified with respect to exact cardinal values (singular, dual, trial). Analog magnitude representations include symbols for quantity that are embedded within a system of arithmetical computations, but they represent only approximate cardinal values; there is no representation of exactly 1, and therefore no representation of 1. Analog magnitude representations cannot even resolve the distinction between 10 and 11 (or any two successive integers beyond its discrimination capacity), and so cannot express the successor function. Therefore, none of the CS1s can represent 10, let alone 342,689,455.

The argument appears to be just this: since CP-knowers understand the meanings of count words, they are capable of representing exact cardinalities; yet none of the relevant individual conceptual systems available to CP-knowers – the ANS, the language faculty, and the parallel individuation system – can by itself represent exact cardinalities; thus, representing exact cardinalities constitutes a distributive discontinuity.

Is this claim correct? Though we grant, for the sake of argument, that neither the ANS nor the parallel individuation system have the resources to represent exact cardinalities, we doubt that her claims about language are correct. Specifically, given Carey's assumptions about what count words denote and what the relevant semantic machinery is like, language alone seemingly provides the resources required to do so.

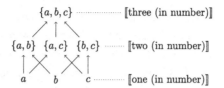

Figure 4 A set-based semantics for cardinality predicates.

Here's why. According to Carey, there is an upper limit of three on the cardinalities "set-based quantification" represents. By **set based quantification**, Carey intends the same set-based semantics for nouns discussed in Section 3.4.2, illustrated in Figure 3. Thus, Carey (2009a, p. 256) writes:

> Linguists such as Genaro Chierchia (1988) and G. Link (1987) show how the quantifiers of all languages, as well as the count/mass distinction, the distinction between nouns in classifier languages and nouns in languages with the count/mass distinction, and much else, can be defined over the semi-lattice depicted in [Figure 3], which creates all of the possible sets from a domain of atoms or individuals. This structure makes explicit the contrast between individuals (the bottom line) and all of the sets composed out of them. One needs explicit symbols with the content [SET] and [INDIVIDUAL], plus distributive and collective computations over those symbols, to capture the meanings of natural language quantifiers, including even the singular/plural distinction. I call this system of representation "set-based quantification."

Apparently, then, set-based quantification includes the entire semi-lattice structure depicted in Figure 3 – the urelements, the sets of urelements, and the relations on them. Moreover, according to Carey, these resources collectively impose "an upper limit on the sets sizes that are quantified with respect to exact cardinal values."

Yet this conclusion strikes us as unfounded. Indeed, on Carey's own assumptions regarding set-based quantification, there is no evident reason why it cannot represent exact cardinalities far greater than three.[24] On standard analyses of numerical modifiers, including that of Link (1983) and Chierchia (1998a,b), cardinality predicates denote collections of precisely the sorts of entities depicted within the relevant semilattice structures. For example, a set-based semantics will naturally provide denotations for cardinality predicates like those depicted in Figure 4.

---

[24] In Section 4, we reconsider Carey's assumption that the relevant semantics is set-theoretic. Specifically, we note the existence of alternative, *mereological* analyses, such as that of Link (1983).

These same denotations are what Carey identifies with *cardinal values* – collections of equinumerous sets.[25] Moreover, Carey clearly identifies cardinal values with *numbers*: "cardinal values of sets are numbers" Carey (2009a, p. 136). Hence, it would appear that set-based quantification alone provides the resources for representing exact cardinal values. Furthermore, it should be evident that there is no clear upper-limit on the cardinal values representable within such a system. Indeed, for any number of urelements, n, the system is capable of representing numbers up to n.

We thus conclude that, by Carey's own lights, there is little reason to posit a conceptual discontinuity for the case of natural number. More precisely, since a single core system – the language faculty – would alone suffice to represent the relevant sets, it follows that there is no distributive discontinuity in the case of natural number. Moreover, since the absence of a distributive discontinuity implies the absence of a cumulative discontinuity, it also follows that there is no cumulative discontinuity. Thus Carey's account of natural number concepts fails to secure conceptual discontinuity in either sense.

4 Assessing the Mainstream

In the preceding sections, we made explicit the primary issues the Mainstream seeks to address, along with various assumptions Mainstream views typically adopt. Additionally, we outlined and critically assessed a range of influential Mainstream accounts of how children come to possess, and manifest competence with, number concepts. In this final section, we evaluate certain additional, widely shared commitments of Mainstream NCR, which should be of particular interest to philosophers of mathematics.

As we saw in Section 2, the Mainstream incurs *many* commitments – to representationalism, to the existence of core systems, to a relatively detailed ontogenetic timetable, and so on. Though each is deserving of critical attention, we focus here on commitments of most direct concern to philosophers of mathematics, largely because, in contrast to many of the Mainstream's other commitments, these have received comparatively less critical scrutiny within NCR. Specifically, we focus on three issues:

- What is the relationship between psychology and foundational theories of arithmetic?
- What is the nature of natural numbers?
- How many kinds of number concepts do we possess?

---

[25] Assuming Quine's trick. See fn. 15.

Bringing these to the fore will, we hope, facilitate progress not only by pinpointing and assessing relatively neglected theoretical assumptions underlying much NCR, but also by highlighting points of contact between otherwise disparate fields of study.

4.1 Psychology and Foundational Mathematics

Debates regarding the relationship between psychology and foundational mathematics are complex and long-standing – too elaborate to discuss in detail here. Still, one important issue of direct relevance to NCR is:

The Constraint Question: To what extent should research within foundational mathematics constrain NCR, or vice versa?

Some clarifications are in order. First, we use 'foundational mathematics' as a loose catchall to include, e.g. logic, arithmetic, set theory, category theory, and metamathematics. Second, though many notions of *constraint* are potentially relevant here, what we principally have in mind is a relationship between research fields such that the range of permissible hypotheses in one is restricted by research in the other. Presumably, every field constrains every other, in this sense, at least to the extent that they should be consistent. However, the constraint relations between some fields are more extensive, in that they impose what we'll call **specific restrictions**. For example, physics imposes greater degrees of restriction on hypotheses in, say, chemistry, than it does in, say, history, since particle physics provides an account of the nature of the elements that figure in the periodic table. Crudely, the Constraint Question is concerned with whether the constraint relations between NCR and foundational mathematics are more akin to those between physics and chemistry than between physics and history.

So construed, there are two familiar extreme positions that may be found in discussions of the Constraint Question:

Radical Anti-Psychologism: NCR imposes no specific restrictions on foundational mathematics.
Radical Psychologism: NCR imposes substantial specific restrictions on foundational mathematics.

Though few would perhaps accept either claim without further clarifications and caveats, we suspect that the first roughly captures a prevailing view among philosophers of mathematics. Moreover, while the second is far less prevalent among contemporary philosophers, it is suggested in different ways by those who view logic and other foundational fields as part of psychology, and by those who claim that mathematical objects are psychological entities. Indeed, in Section 4.2 we discuss Mainstream views that appear to make this latter claim. Here, we focus instead on an answer to the Constraint Question explicitly defended by Susan Carey, but which we suspect is more widely accepted among NCRs.

Carey (2009b) proposes a midway between the aforementioned pair of extremes, which focuses on the role foundational research plays in constraining NCR. On her view, foundational results provide a *specification* of the number concepts acquired throughout the course of development. That is, foundational mathematics specifies the salient *developmental target*.

Carey (2009b, pp. 1–2) begins by distinguishing between two kinds of programs: **the logical program** and **the ontogenetic program**:

> Two distinct research programs should be, but often are not, distinguished. In one, the logical program, the building blocks are conceived of as logically necessary prerequisites for the capacity in question. In the case of natural number representations these might include the capacity for carrying out recursive computations, the capacity to represent sets, and various logical capacities, such as those captured in second order predicate calculus. ... A second research program, the ontogenetic program, conceives of the building blocks as specific representational systems out of which the target representational capacity is actually built in the course of ontogenesis or historical development. In the case of number representations these would be the innate representations with numerical content (if any).

The two programs are related, but distinct, Carey explains:

> The first (characterizing the logical prerequisites for natural number) leads to analyses like those that attempt to derive the Peano–Dedekind axioms from Zermelo–Fraenkel set theory or Frege's proof that attempts to derive these axioms from second order logic and the principle that if two sets can be put in 1-1 correspondence they have the same cardinal value. Such analyses seek to uncover the structure of the concept of natural number, and certainly involve representational capacities drawn upon in mature mathematical thought, but nobody would suppose that in ontogenesis or historical development people construct the concept of natural number by recapitulating such proofs.

Despite this, and crucially, the ontogenetic program depends on the logical program, in two respects:

> The ontogenetic project ... depends upon the logical one for a characterization of the target concept at issue (e.g., what I mean here by the concept of natural number is characterized by the Peano–Dedekind axioms), as well as for a characterization of the logical resources drawn upon in the construction process.

Thus, Carey's view seemingly implies two commitments. First, successfully carrying out the logical program is a *prerequisite* for successfully carrying out the ontogenetic program, since the former provides not only a characterization of the target concept(s), but also a specification of the logical resources required to possess them. Because these are the purview of foundational theories, developmentalists have no choice but to turn to such enterprises when attempting to specify the developmental target. To give it a label, call this methodological orientation **Foundations First** (FF).

A second, perhaps less obvious implication of Carey's view concerns the adoption of what we call **conceptual monism**. Specifically, Carey's framing of the relationship between foundational mathematics and NCR seemingly implies that foundational research delivers a *unique* target concept – "*the* concept of natural number [as] characterized by the Peano–Dedekind axioms" (our emphasis). More generally, the presumption appears to be that each number word expresses a unique number concept – 'one' expresses ONE, 'two' expresses TWO, and so on – and that possessing these subordinate concepts suffices for possessing NATURAL NUMBER. We will return to this in Section 4.3. For now, we focus on the tenability of FF.

As we see it, FF faces three kinds of challenges. The first is to justify the methodology, in light of its failing to generalize to other conceptual domains. A basic assumption underlying FF appears to be that in order to specify a target number concept, we ought to look to those sciences most relevant to understanding the target phenomenon: in this case, foundational mathematics. Now, consider the concept WATER. Though it may be useful for developmentalists to know how chemists conceptualize water, this surely isn't *required* to specify the target concept, i.e. the one acquired by children when learning to think and talk about water. Presumably, psychology and cognitive science are not, and should not be, hostage to chemistry in this way. But then why should things be interestingly different with respect to foundational mathematics?

Second, in order for foundational theories to play the role assigned to them by FF, they would need to accurately characterize the concepts children in fact acquire, i.e. the developmental target. However, foundational theories are usually not developed with an eye to satisfying such a demand.

For one thing, and contrary to what Carey seemingly claims, it is doubtful that foundational theories of the natural numbers, such as the DP-axioms, "seek to uncover the structure of the concept of natural number" – not at any rate if concepts are mental particulars which are constituents of quotidian thoughts. Rather, such axioms are more plausibly construed as concerning *objects* and the structure they instantiate, viz. natural numbers.

Furthermore, even if we construe foundational theories as seeking to provide theories of mathematical concepts, as opposed to mathematical entities or
structures, it is important to appreciate that such theories are often intended to
be *revisionary*, rather than descriptive. Appealing again to Strawson's (2002)
distinction, foundational theories would play the role Carey assumes only if
they were engaged in *descriptive* metaphysics – the entities and structures posited aim to describe how we actually think (Section 1.2). Yet they are very
typically not. To take one well-known example, consider Frege's (1884 §57)
analysis of number, whose explicit aim is not to accurately describe our quotidian number concepts, but rather to *improve* them: "to define a concept of
number that is useful for science." Thus, even if we suppose that foundational
theories are theories of mathematical concepts, there's little reason to suppose
that they seek to characterize the quotidian concepts we in fact possess.

A third challenge concerns adjudicating between competing foundational
theories. As Rips et al. (2008) observe, Carey, like most NCRs, seemingly
assumes that foundational mathematics delivers a uniquely correct answer to
the question of what kinds of things natural numbers are, namely *cardinalities*, thanks to the foundational work of, e.g. Frege (1884) and Russell (1919).
However, this is only one among *many* options. Thus, Snyder et al. (2018b)
distinguish three kinds of characterizations of the naturals, each plausibly codifying a different potential application: counting, ordering, and calculating.
These are:

Cardinal Characterizations: Natural numbers are cardinalities, i.e. the sorts
   of things answering 'how many'-questions. Cardinalities might be identified
   with properties or classes of equinumerous concepts [Frege (1884); Russell
   (1919); Hale and Wright (2001)], or sets [Maddy (1990); or tropes (Moltmann
   (2013))] or they might instead be characterized as "sui generis logical objects"
   [Tennant (1997)].
Ordinal Characterizations: Natural numbers are **ordinalities**, i.e. the sorts of
   things specifying the ordinal positions of objects within a linearly ordered
   sequence. They might be identified with order-types of (finite) well-ordered
   sets [Dummett (1991a)], or equivalence classes of numeral-ordering pairs
   [Linnebo (2009)].
Structuralist Characterizations: Natural numbers are reified positions or
   places within an ω-**sequence**, i.e. the kind of sequence characterized by the
   Dedekind-Peano axioms [Resnik (1997); Shapiro (1997)].

Given the availability of so many characterizations, any particular choice
clearly demands some justification. However, this is very rarely offered by

Mainstream NCRs. In our view, this highlights a significant opportunity for interdisciplinary engagement between NCR and the philosophy of mathematics. After all, deciding how best to characterize the naturals is both a long-standing preoccupation of the latter, and may very well impact how empirical hypotheses are framed within NCR. For example, it may directly impact assumptions regarding what is required to possess natural number concepts [Rips et al. (2008)].

4.2 Metaphysics of Natural Number

The following question is of central interest to the philosophy of mathematics, of course:

The Metaphysical Question: Assuming natural numbers exist, what are they like?

However, given the empirical focus of NCR, one might reasonably expect it to incur no commitments regarding such matters. It may come as some surprise, then, that some Mainstream NCRs routinely adopt controversial metaphysical positions concerning the natural numbers. In what follows, we consider three: cultural constructionism, term formalism, and the Frege–Russell characterization.

4.2.1 Cultural Constructionism

Cultural constructionism is the thesis that numbers depend for their existence on quite specific cognitively mediated cultural practices. Dehaene (2011, p. 260) seemingly adopts this view when writing:

> The great mathematician Leopold Kronecker was wrong when he claimed that "God made the integers; all else is the work of man." Even the integers are manmade. They only exist in cultures that invented the notion of counting.

In a similar vein, Carey (2009a, p. 333) writes:

> Kronecker was wrong. Neither God nor evolution gave humans natural number. Natural number is a human construction.

To avoid confusion, Carey and Dehaene's cultural constructionism is not intended to be a **constructivism** (or **intuitionism**) of the sort normally associated with Brouwer or Heyting.[26] Nor is it a constructivism about psychological

---

[26] See Shapiro (2000, Ch. 7).

processes of the sort associated with Piaget. Rather, they appear to be endorsing a kind of *ontological dependency thesis* regarding the natural numbers. In a slogan: no natural numbers without human cognition (of the relevant sort). As such, numbers are not *discovered* by human beings, but rather *invented*. Specifically, both Dehaene and Carey seemingly suppose that this invention – this bringing into being – depends on the existence of counting procedures.

What should we make of this view? Given its distance from Carey and Dehaene's primary intellectual concerns, we're inclined to think of its endorsement is a kind of rhetorical flourish – something not intended to be central to the overall views being defended.[27] For one thing, the aforementioned quotations occur in the absence of any serious argumentation. In Dehaene's case, for example, the claim that "the integers are manmade" occurs within the context of claiming that "what occurs in the child's mind when he suddenly understands that there is a discrete infinity of exact numbers" depends on "a *cultural* invention." However, it is clearly one thing for the child's *understanding* of natural numbers to rely upon cultural innovations, quite another to claim that *numbers* are cultural inventions. To suppose otherwise involves something akin to a use-mention confusion – to conflate the origin of numbers with that of mental states representing numbers.

Regardless, there are independent, well-known reasons for rejecting cultural constructionism. Specifically, following familiar arguments from Frege, it seems incapable of accommodating two widely assumed properties of mathematical truths:[28]

- *Eternality*: Mathematical truths were true before humans existed, and will remain true if humans cease existing. This is why it is permissible to apply mathematics not just to the present and future, but also to the distant past.
- *Necessity*: Mathematical truths are necessary, rather than contingent. Thus, in all conceivable ways the world could be, including those in which no cultural practices exist, 3 + 2 would be equal to 5. This is why it is permissible to apply mathematics to both actual and counterfactual circumstances.

But if numbers only came to exist once humans engaged in counting behavior, then it is hard to see how arithmetic truths could have either property. After all, counting procedures might never have been invented, and mathematical truths presumably held prior to their invention. Altogether, then, cultural constructionism is rather implausible.[29]

---

[27] Though this might be overly generous. After all, as a reviewer observes, the subtitle to Dehaene's books is 'How the Mind Creates Mathematics'.
[28] Cf. Shapiro (2000) and Linnebo (2017).
[29] Though see Cole (2013) for an attempt to address similar kinds of problems.

4.2.2 Term Formalism

A second familiar metaphysical thesis about numbers, sometimes invoked by NCRs, is **term formalism**. Crudely, this is the view that natural numbers are symbols we use to talk about or represent them. Thus, (17) is true if 'two', which refers to itself, is an even number.

(17)　Two is an even number.

Sometimes term formalism is explicitly endorsed in NCR. For example, Gallistel (2021, p. 29) recently writes: "Numbers are symbols manipulated in accord with the axioms of arithmetic." He then proceeds to summarize the earlier view of Gelman and Gallistel (1986) – itself a seminal text in NCR – which endorses "the modern formalist view of arithmetic":

> A number, as we understand it, is a player in the game of arithmetic, defined by the rules of that game.

Although Gelman and Galistel's commitment to term formalism is rather explicit, it can also be found, though perhaps less overtly, in other NCR. For example, the preface to Dehaene (2011, p. xiii) begins:

> We are surrounded by numbers. Etched on credit cards or engraved on coins, printed on pay checks or aligned on computerized spreadsheets, numbers rule our lives.

Similarly, while speaking of CP-knowers, Davidson et al. (2012, pp. 3–4) suggest that items comprising a count list are numbers.

> [A]lthough these children can recite higher numbers (e.g., 5 or 10), and know that these higher words contrast in meaning… they appear to lack meanings for the rest of the words in their count list.

The NCR literature is replete with similar comments. However, such pronouncements only make sense if numbers are *symbols*: if what is etched on credit cards, engraved on coins, or recited are themselves *numbers*.

Term formalism is well-known within philosophy primarily because of what Weir (2022) colorfully describes as "a demolition job by a great philosopher, Gotlob Frege." Indeed, Frege's (1903) polemic against term formalism has long been widely regarded as decisive. Here, we list a couple well-known problems, also due to Frege, as articulated by Linnebo (2017):[30]

---

[30]　See also Shapiro (2000).

- *Identity*: According to term formalism, an arithmetic identity statement such as '3 + 2 = 5' (or 'Three plus two equals five') is true just in case '3 + 2' is coreferential with '5'. However, assuming both terms refer to themselves, this is false. Thus, term formalism is committed to an empirically unsupported ambiguity in the equative copula: it expresses the familiar notion of identity when flanked by nonmathematical terms, e.g. 'Superman' and 'Clark Kent', and an altogether different notion when featuring mathematical terms.

- *Infinity*: Initially, the philosophical motivation for adopting term formalism is metaphysical and epistemological. Specifically, how can we know that 3 + 2 = 5 if numbers are *abstract* [Benacerraf (1973)]? Thus, it is tempting to identify numbers with concretia, in the form of linguistic tokens, e.g. particular inscriptions of, e.g. '3 + 2' and '5'. But given the truth of universally quantified statements like 'Every natural number has a successor', infinitely many numbers are required. Yet we have no good reason to suppose that infinitely many concrete linguistic tokens exist. Indeed, as Frege (1884, §123) quips: "We have neither an infinite blackboard, nor infinitely much chalk at our disposal." Hence, the only plausible form of term formalism is one which posits infinitely many expression **types**, and types are presumably *abstract*. Hence, the purported epistemic benefit of adopting term formalism – direct perceptual access to mathematical objects – is lost.

Again, such considerations render term formalism rather implausible.

4.2.3 The Frege–Russell Characterization

A third metaphysical thesis, which we encountered in Section 1.1 and again in Section 3.5, is known as *the Frege–Russell characterization*, because of its association with Frege (1884) and Russell (1919). According to it, natural numbers are collections of equinumerous, or bijective, sets. For example, the number two is just the collection of all two member sets. Again, as Rips et al. (2008) point out, this appears to the be the predominant view among NCRs. For example, as we saw in Section 3.5.2, Carey (2009a, p. 136) adopts this view when identifying numbers with "cardinal values." Similarly, vanMarle (2018, p. 131) writes: "As described by Frege (Frege, 1884/1980), formally, numbers are a special kind of sets."

Within NCR, the Frege–Russell characterization is often adopted as a kind of metasemantic thesis. Specifically, NCRs often claim that count words or concepts denote collections of sets, and thus natural numbers. For example, speaking of the denotation of 'two', vanMarle (2018, p. 131) writes:

> [I]n order to learn what 'two' means, the child must abstract away from the immediate perceptual features of the items in a set (color, size, shape, kind, etc.) in order to discover that 'two' refers to any set made up of exactly two individuals.

Expanding on this, Bloom and Wynn (1997, p. 512) write.

> The word *two* in the phrase *two black cats* does not describe any individual in the external world, nor does it refer to a property that any individual in the world might possess. In this regard, it differs from the noun *cats*, which is understood as describing cats, or the adjective *black*, which is understood as describing a property that each of the individual cats might have. Rather *two* is a predicate that applies to the set of cats. More generally, as Frege (1893/1980) has argued, numbers are predicates of sets of individuals.

The reference here is to Frege's (1884, §46) well-known contention that "statements of number," such as (18), are not really about objects, e.g. Elmos, but rather Fregean concepts, such as being two in number.

(18)   Those are two Elmos.

Quoting Frege (1884, §46) directly:

> It should throw some light on the matter to consider number in the context of a judgment which brings out its basic use. While looking at one and the same external phenomenon, I can say with equal truth both "It is a copse" and "It is five trees," or both "Here are four companies" and "Here are 500 men"... If I say "The King's carriage is drawn by four horses," then I assign the number four to the concept "horse that draws the King's carriage."

Speaking in Fregean terms, Frege's claim is that to assert (18) is to assert that a certain first-level concept – 'is an Elmo' — falls under a certain second-level concept — 'is two (in number)'. Given the Frege–Russell characterization, this second-level concept *just is* the number two.

In more contemporary semantic parlance, second-level concepts are *generalized quantifiers* (Section 3.4.2): second-order predicates which take first-order predicates as arguments and return a truth-value. Prototypical examples include the underlined expressions in (19).

(19)   {Every person/Nothing/Some number} is interesting.

Within **Generalized Quantifier Theory** [Barwise and Cooper (1981)], these take the predicate 'is interesting' as argument, and return True just in case the set of interesting things is a superset of the people, or is disjoint with everything, or overlaps with the numbers, respectively. In the previous quote, Bloom and

Wynn suggest that 'two' in 'two black cats' belongs to this same category: it is a second-order predicate true of the first-order predicate 'is a cat'. Furthermore, and crucially, it denotes a *number*.

As we saw in Section 4.1, as a metaphysical thesis, the Frege–Russell characterization is only one of several formal characterizations of the naturals. Our primary concern here is with the implications of this thesis for the issue of what number words denote. If the Frege–Russell characterization were correct, then such entities should presumably be the referents of number words used referentially – i.e. *numerals*. But contrary to what many NCRs suppose, there are good reasons to doubt that such uses of number words denote these entities.

The first objection targets the semantic function of number words. Number words have nonreferential uses, as witnessed by 'two' in (18), as well as referential uses, as witnessed by 'two' in (20).

(20)   Two is an even number.

Within contemporary semantic theory, these different occurrences have different semantic types [Snyder (2021b)]. Moreover, what an expression denotes is determined by its semantic type, so that if different occurrences of the same expression have different semantic types, then they must also have different denotations. It follows that the different occurrences of 'two' in (18) and (20) cannot be co-denoting. In particular, if 'two' in (18) denotes a collection of sets, then, contrary to fact, 'two' in (20) would not refer to a number – not, at any rate, if that number *just is* that collection of sets.

In more detail, we can formulate the problem based on a principle which is widely assumed within linguistic semantics, namely:

Denotation-to-Type: For any two occurrences o_1 and o_2 of an expression α, o_1 and o_2 have the same denotation only if o_1 and o_2 have the same semantic type.

This reflects the usual explanation of compositionality – the semantic function of an expression is determined by its semantic type, and semantic types combine so as to ensure that the denotation of the whole is a function of the denotations of the parts. To illustrate, consider (21).

(21)   a. The Elmos on the table are red (in hue).

        b. Those are red Elmos (on the table).

        c. Red is a primary color.

On standard accounts, the different occurrences of 'red' here have different semantic types, and thus different denotations [McNally et al. (2011)]. Specifically, in (21a), 'red' has the type of a predicate, denoting red things; in (21b), 'red' has the type of a modifier, denoting subsets of red things, in this case Elmos on the table; and in (21c), 'red' has the type of a name, referring to the color red, as an entity.

Of course, as we've seen, number words exhibit a similar pattern of uses [Snyder et al. (2022)]. Specifically, while 'two' in (20) has the type of a name, referring to a number, as an entity, 'two' in (1a) has the type of a predicate, denoting pluralities of two things, and 'two' in (18) has the type of a modifier, denoting pluralities which are both two in number and, in this case, Elmos on the table.

(1a)   The Elmos on the table are two in number.

Now, here's the argument. The different occurrences of 'two' above have different semantic types. Specifically, whereas 'two' in (20) has the type of a name, 'two' in (18) has the type of a modifier, which according to Bloom and Wynn is that of a generalized quantifier. Thus, by Denotation-to-Type, these different occurrences cannot be co-denoting, contrary to what would be required if numbers were collections of equinumerous sets.

A second problem targets what we call

The Identity Thesis: Natural numbers are cardinalities.

As mentioned, the Frege–Russell characterization is a *cardinal* characterization of the naturals. The present problem poses a challenge for *all* such characterizations: the Identity Thesis generates numerous false predictions.

To see why, consider Frege's (1884) highly influential analysis, whereby all expressions underlined in (22a-c) are coreferential singular terms, referring to the same number.

(22)   a. Two is an even number.

        b. The number two is even.

        c. The number of Elmos (on the table) is two.

Yet despite its considerable influence, there is ample evidence that as a proposal about the semantics of natural language expressions, Frege's identification of the natural numbers with finite cardinalities is mistaken. Specifically, the evidence suggests that natural languages distinguish the sorts of things referenced by singular terms like those in (22a,b) – *numbers* – from those apparently referenced by terms like those in (22c) – *cardinalities*.

Specifically, following Moltmann (2013), let's call phrases of the form 'the number of Elmos' **the number of-terms**, and those like 'the number two' **explicit number-referring terms**. A wealth of semantic contrasts, due mostly to Moltmann, reveal that these are not plausibly coreferential. One kind of data concerns predicates like 'count', 'compare', and 'surprise'. As (23a-d) suggest, even if Mary happened to count, compare, or be surprised by two Elmos, it does not follow that she counted, compared, or was surprised by *an abstract arithmetic object*.

(23) a. Mary counted the number {of Elmos/??two}.

b. Mary compared the number {of Elmos/??two} to the number of men.

c. Mary was surprised by the number {of Elmos/??two}.

d. The bridge collapsed because of the number {of cars/??200}.

Yet this is not to be expected if 'the number of Elmos' and 'the number two' both refer to the number two. We see parallel contrasts in (24a,b).

(24) a. The number of Elmos on the table is {two/??the number two}.

b. The number Mary is writing about is {two/the number two}.

On Frege's analysis, (24a,b) are both identity statements, equating the referents of coreferential singular terms. But since we generally expect coreferential expressions to be acceptably substitutable within identity statements, it is difficult to see how there could be a difference in acceptability in (24a) without there being an analogous difference in (24b). Worse, we see a similar contrast with overt questions [Snyder (2017)].

(25) a. How many Elmos are on the table? {Two/??The number two}.

b. Which number is Mary writing about? {Two/The number two}.

Cardinalities, recall, characteristically answer 'how many'-questions. Thus, the fact that 'the number two' is not acceptable in (25a) suggests that it cannot refer to a cardinality.[31]

Based on these and similar contrasts, Moltmann (2013) and Snyder (2017) both conclude that natural language draws an ontological distinction between numbers and cardinalities. For our purposes, however, the crucial point is that the Identity Thesis, when combined with other independently plausible

---

[31] As a reviewer notes, 'the cardinality two' is also unacceptable in response to (25a). However, this is plausibly for independent reasons. Unlike with 'number', we cannot use 'cardinality' as a monadic predicate applicable to numbers: 'Two is a (prime) {number/??cardinality}'. Thus, we cannot shift 'cardinality' to the semantic type of a modifier: 'the (prime) {number/??cardinality} two'. See Snyder (2017, 2021b).

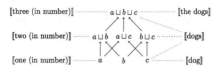

Figure 5 A mereology-based semantics for nouns.

assumptions, implies numerous false predictions. Thus, we have excellent reason for rejecting it.

So far we have argued on semantic grounds that the Frege–Russell characterization is not plausible as an account of what natural numbers are. Still, it might be thought that only a modest refinement of what NCR is about is needed. Specifically, since NCR is principally concerned with our abilities to determine and distinguish *cardinalities*, one might plausibly suggest that cardinalities, not natural numbers, are the primary denotata of early numerical cognition. Thus, one might propose the Frege–Russell characterization as an adequate account of *those* entities. However, we conclude this section by sketching reasons for doubting even this position. Specifically, there are reasons to doubt that the denotations of number words and number concepts should be characterized set-theoretically, at all.

Before considering our concerns regarding the aforementioned position, a terminological observation is in order. Within NCR, 'set' is used in two importantly different senses. First, there's an informal sense, roughly synonymous with 'collection', 'plurality', or 'group', often used when discussing various cognitive systems, such as the ANS or the SNS, which are said to track or represents "sets" in this sense. However, NCRs also often use the mathematical sense of 'set', applicable to entities described by set theory. As emphasized in Section 3, the latter is often adopted when characterizing NUMBER and various concepts purportedly required to possess it, such as EQUINUMEROSITY or SUCCESSOR. Moreover, it is this latter mathematical sense of 'set' that it is used in the Frege–Russell characterization of natural number.

In light of this potential equivocation, it bears emphasizing that while NCRs often suppose that representing cardinality requires possessing mathematical representations, such as SET, ELEMENT, and so on, this is not the *only* viable option. In fact, cardinality is just as readily represented within mereological frameworks, such as that Link (1983), illustrated in Figure 5.

Indeed, most extant analyses of number expressions assume precisely such a framework [Snyder (2021b)]. Hence, it is entirely possible that understanding the meanings of such expressions depends only on the possession of mereological representations – e.g. ATOM, SUM, and PARTHOOD.[32]

---

[32] See Yi (2018) for yet another alternative, based on plural logic.

A second well-known problem concerns perceptual access. The following thesis appears widely assumed within NCR:

Cardinality Perceivability: Some cardinalities are perceivable.

The likely assumption is that without Cardinality Perceivability, it would be hard to explain how we are can perceptually determine how many things are in a collection, or that one collection is larger than another. However, as Yi (2018) observes, combining this thesis with a set-theoretic characterization of cardinalities would appear to require also endorsing

Set Perceivability: Some sets are perceivable.

Indeed, some NCRs appear to explicitly endorse such a thesis. For example, vanMarle (2018, p. 138) writes: "[Children] are assumed to be literally building their understanding of number through repeated experiences seeing the creation of sets and instances of counting." Yet *seeing* sets clearly presupposes Set Perceivability. For our purposes, the important observation is that since abstract objects cannot be objects of perception, presumably, Set Perceivability entails an additional, highly controversial claim:

Impurity: Impure sets whose members are concrete are themselves concrete.[33]

Despite being adopted by some mathematical empiricists,[34] this thesis is also typically rejected by those who suppose, quite generally, that sets are *abstract* [Yi (2018)]. In contrast, given traditional assumptions, adopting a mereological analysis invites no such controversy. Specifically, mereological sums of concretia are traditionally assumed to be concrete. Indeed, the avoidance of abstracta was among the original motivations for developing mereology [Goodman and Quine (1947)].

4.3 Conceptual Pluralism

In Section 4.2, we noted that number words have different kinds of uses. Here we argue from this observation to the conclusion that number concepts are subject to two different sorts of pluralism: what we call 'synchronic pluralism' and 'diachronic pluralism'. Before doing so, however, we draw out some important implications concerning the semantics of number words.

---

[33] A set is "impure" if it contains urelements, i.e. members which are not sets.
[34] See, e.g. Kim (1981) and Maddy (1990).

We have already seen that number words have different kinds of uses, and we will see next that they have *many* more. For example, they can function as names referring to numbers (as entities), and they can function as predicates predicating cardinality properties. Yet 'two' is not **lexically ambiguous** in the way that, e.g. 'bank' is. That is, whereas different uses of 'bank' express different, completely unrelated meanings (one applicable to financial institutions, another to river banks), on standard accounts within linguistic semantics, number words are **polysemous**, having a wide variety of different, though *related*, meanings [Snyder (2021b)]. Thus, we have a kind of *pluralism* with respect to the meanings of number words: there are multiple, related meanings expressed by different occurrences of the same word. However, we maintain that this also supports a similar kind of pluralism with respect to number *concepts*.

4.3.1 Synchronic Pluralism

The first kind of pluralism we defend concerns number concepts as possessed by adults having linguistic competence with number words:

Synchronic Pluralism: Quotidian adult numerical thought is such that for each number word in a natural language, there are multiple corresponding number concepts.

As we see it, this is a consequence of theses that are either argued for above, or else widely assumed within NCR. The first is

Polysemy: Different occurrences of the same number word in different syntactic environments express different, related meanings.

We have already seen that number words, such as 'two', take on different semantic functions, and thus denotations, in different syntactic environments. Furthermore, in some contexts, different occurrences of number words can serve the same semantic function – referring – while plausibly denoting different sorts of things – numbers and cardinalities. However, this is merely scratching the surface. Here's an expanded, yet still nonexhaustive, list of how 'two' can be used in different contexts [Snyder (2021b)]:

(26) a. Mars' moons are <u>two</u> (in number).   (predicative)

b. Those are (Mars') <u>two</u> moons.   (attributive)

c. Mars has <u>two</u> moons.   (quantificational)

d. Mars' moons number <u>two</u> (in total).   (verbal complement)

e. The number of Mars' moons is <u>two</u>.   (specificational)

f. <u>Two</u> is an even number.   (numeral)

g. The number <u>two</u> is even.   (predicative numeral)

h. Michael Jordan is roughly <u>two</u> meters tall.   (measurement)

i. Rafael Nadal is ranked number <u>two</u> in the world.   (ordinal)

The labels here indicate how 'two' is being used in the accompanying example, i.e. how it's functioning semantically. For example, 'two' in (26a) is used predicatively, since it functions as a predicate, similar to 'large' in (27a). Similarly, 'two' in (26b) is used attributively, since it functions as an attributive adjective, similar to 'large' in (27b).

(27)   a. Mars' moons are <u>large</u> (in size).

    b. Those are (Mars') <u>large</u> moons.

Since an expression's function is determined by its semantic type in a given syntactic context [Partee (1986)], many of the occurrences of 'two' in (26) have different semantic types. Moreover, since the meaning of an expression is also a function of its semantic type, (26) is witness to the fact that 'two' is (radically) *polysemous* [Snyder (2021b)].

In light of Polysemy, there is a clear route to Synchronic Pluralism. First, it is widely assumed that understanding a natural language expression requires the possession of a corresponding concept, having the same content (Section 2.2). For example, understanding (26a), where 'two' is used to predicate a cardinality property, requires possessing a concept having the same content, i.e. one representing a cardinality property. However, on standard assumptions, concepts are at least partially individuated by their contents. Thus, the concept required to understand 'two' in (26a) is distinct from the one required to understand 'two' in, say, (26f), where 'two' is being used to refer. Indeed, the concepts differ not just in their contents, but also their representational functions: one functions predicatively, the other referentially.

Thus, rather than supposing that there is a single concept, TWO, corresponding the word 'two', plausibly there are at least two concepts corresponding to the word 'two':

TWO$_{num-ref}$: A concept referring to the number two, as an entity. Call concepts of this sort **numerical concepts**.

TWO$_{card-pred}$: A concept predicating the property of being two in number of collections. Call concepts of this sort **cardinal concepts**.

Indeed, though we won't argue for this here, we think that understanding the various uses of 'two' in (26) plausibly requires the possession of even more concepts, including, but not limited to:

TWO$_{card\text{-}ref}$: A concept referring to the cardinality property of being two in number, as an entity.
TWO$_{ord}$: A concept specifying the ordinal position of an object among a linearly ordered class of objects.

In any case, it would appear that the polysemy of number words, along with standard representationalist commitments, jointly entail a commitment to Synchronic Pluralism.

4.3.2 The Complex Transition Thesis

Much of developmental NCR focuses on number cognition up to the age of around $3\frac{1}{2}$ or 4, when children become CP-Knowers. However, development doesn't end here, of course. In particular, NCR must also explain the subsequent development of arithmetic abilities, from approximately five to eight years, when children are exposed to basic **symbolic arithmetic**, and learn to manipulate numerical symbols, such as Arabic numerals, to solve arithmetic problems (Section 2.4).

One obvious question concerning this period of development is whether it requires the acquisition of novel concepts, or instead whether those concepts possessed through learning to count suffice. Roughly put, are the concepts CP-knowers possess in virtue of understanding count words – e.g. 'one', 'two', and 'three' – the same as those required to interpret symbolic numerals in basic arithmetic equations – e.g. '1 + 2 = 3'?

One potential answer here is

Simple Transition: Early CP-knowers already possess those number concepts required to understand symbolic numerals.

On such a view, learning symbolic arithmetic does not require the acquisition of novel concepts. Instead, it involves enhancing the child's abilities to deploy the same concepts already possessed at $3\frac{1}{2}$–4 years of age.

To be clear, Simple Transition is "simple" in that it minimizes the stock of concepts required to successfully do elementary arithmetic. Specifically, because the concept required to understand, say, the count word 'two' is the same as the one required to understand the symbolic numeral '2', only one collection of number-related concepts is needed to explain the transition from counting to elementary arithmetic. Below, we will entertain a competing hypothesis, which is "complex" in virtue of positing *distinct* collections of number-related concepts – one for understanding count words, and one for understanding symbolic numerals.

For now, suppose Simple Transition is correct, in which case understanding symbolic numerals only requires possessing count concepts. But what are these concepts, exactly? In light of the empirical evidence, the concepts most plausibly attributed to early counters are those like $TWO_{card\text{-}pred}$, which involve ascribing cardinality properties to collections. To see why, consider the sorts of achievements typically taken as evidence of concept possession. These involve *cardinality* – e.g. successfully answering 'how many'-questions, or handing over a specific number of items. Thus, such studies do not provide evidence for the possession of $TWO_{num\text{-}ref}$ – a concept *referring* to the number two. Rather, they more plausibly provide evidence for $TWO_{card\text{-}pred}$ – a concept predicating a cardinality *property*.

Thus, under present assumptions, learning basic arithmetic will only require concepts used for representing cardinality. Indeed, quoting Butterworth (2005, p. 3):

> The development of arithmetic can be seen in terms of an increasingly sophisticated understanding of [cardinality] and its implications, and in increasing skill in manipulating [cardinalities].

The suggestion appears to be that the child associates count concepts – those already possessed in virtue of understanding count words – with their symbolic numeral counterparts while associating similar kinds of cardinal meanings with basic arithmetic operations and relations, such as '+' and '=' (Section 2.4). Mastering basic arithmetic is thus not a matter of acquiring new concepts, but rather learning increasingly sophisticated ways of manipulating what the child is already capable of representing – cardinalities.

Simple Transition has much to recommend it. Specifically, in the absence of countervailing evidence, it provides a simple and elegant account of the transition between counting and more sophisticated numerical competences. However, we maintain that there are good reasons to reject it. As Frege (1884) observed long ago, symbolic numerals, as they occur in arithmetic equations like (28), represent numbers *as objects*.

(28) $3 + 2 = 5$

Furthermore, these same equations can be equivalently paraphrased in English using nonsymbolic numerals. For example:

(29)   Three plus two is five.

Yet the thought expressed by (29) most plausibly requires possession of concepts which refer to numbers *as objects*, e.g. $TWO_{num-ref}$, not concepts predicating cardinality properties to collections (Section 4.2.3). Rather, understanding arithmetic equations most plausibly requires possessing number concepts beyond those plausibly attributed to early CP-Knowers.

Thus, we have reason to instead endorse

Complex Transition: Understanding symbolic numerals requires number concepts not already possessed by early CP-Knowers.

In view of this, developmental NCR must confront a challenge that, heretofore, has not been widely appreciated. To explain the development of number cognition, it will be necessary to explain how children acquire concepts that refer to numbers – e.g. $TWO_{num-ref}$ – as well as those ascribing cardinality properties to collections – e.g. $TWO_{card-pred}$.

4.3.3 Diachronic Pluralism

Our view is that taken altogether, the best available empirical evidence supports

Diachronic Pluralism: At different stages in development, human beings possess different number concepts required to understand the same number word.

In fact, we distinguish, and accept, two forms of Diachronic Pluralism. The first is

Weak Diachronic Pluralism: At different developmental stages, humans possess different concepts required to understand different uses of the same number word.

This follows from claims argued for above. Specifically, CP-Knowers plausibly possess concepts that ascribe cardinalities, but lack concepts that refer to natural numbers. Yet at some subsequent developmental stage, CP-knowers come to possess *both* kinds of concepts. For example, quotidian adult thought requires concepts for understanding both predicative and referential uses of 'two'.

The second, perhaps more radical form of Diachronic Pluralism we accept is

Strong Diachronic Pluralism: At different developmental stages, humans possess different concepts required to understand the same use of the same
number word.

Specifically, we accept a form of Strong Diachronic Pluralism on which
understanding different utterances of (1a), at different stages of development,
involves different kinds of cardinal concepts.

(1a) The Elmos on the table are two in number.

Obviously, since 'two' serves the same (predicative) semantic function in different utterances of (1a), we are committed to there being distinct numerical
concepts deployed in understanding this same use of 'two', i.e. different token
utterances of the same utterance type.

Why accept this version of Strong Diachronic Pluralism? In our view, that's
because it represents the best solution to an important, albeit largely unappreciated, puzzle resulting from attempting to align our best developmental
psychology with our best semantics for number expressions. Specifically, it
results from two independently plausible, but jointly inconsistent, theses. The
first is supported by research in developmental psychology:

Developmental Priority: CP-knowers possess cardinal concepts prior to
numerical concepts.

As we've argued, whereas early CP-knowers possess cardinal concepts, we
have reason to suppose that they don't yet possess numerical concepts. After all,
while they are able to understand how number words are used to count, there's
very little reason to suppose that they also understand how number words are
used to refer to numbers, as objects. Certainly, no such ability appears manifest
in their behavior.

The second thesis results from combining independently plausible assumptions about word comprehension with what is, in our view at least, the most
plausible available semantics for number words:

Semantic Priority: One cannot understand cardinal meanings of number words
without also possessing corresponding numerical concepts.

This follows from three assumptions, each of which enjoys a great deal of initial
plausibility. The first is

Meaning Comprehension: Understanding the meaning of a word in a given
context requires possessing a concept having the same content as expressed
by that word in that context.

To illustrate, according to Meaning Comprehension, understanding (30a) requires possessing a concept DOG, which has the same content as expressed by 'dog'. Similarly, understanding (30b) presumably first requires resolving an ambiguity in 'dish' – is it intended to denote pottery or food? – and secondly invoking a concept having the same content as the meaning most likely intended in the relevant context.

(30)   a. Rover is a dog.

   b. Mary created a new dish.

Meaning Comprehension appears to be widely assumed within Mainstream developmental psychology. Indeed, given representationalism, it seemingly provides a natural, if not compulsory, explanation for how we can bear the attitude of understanding towards word meanings.

The second assumption required is

Polymorphic Comprehension: Understanding any meaning of a polymorphic expression requires understanding its lexical meaning.

According to our best extant analyses, number words are polysemous in virtue of being **polymorphic**, taking on a wide range of different semantic types in different syntactic environments, thanks to a semantic phenomenon known as **type-shifting** [Snyder (2021b)]. In general, the various meanings a polymorphic expression can take on are all systematically related, in virtue of being either witness to, or else generated from, its lexical meaning, via type-shifting [Partee (1986)]. For example, the meanings of 'two' noted in (26) are all systematically related in this way, thus explaining its polysemy. So, on the seemingly plausible assumption that one comes to represent these meanings via type-shifting, it stands to reason that one understands *any* meaning of 'two' only if one understands its *lexical* meaning specifically.

The final assumption required is

Substantivalism: The lexical meaning of a number word is that of a numeral, referring to a number.

Although there are different kinds of polymorphic analyses of number words available, in our view the most empirically plausible is what Snyder (2021b); Snyder et al. (2022) call **substantivalism**, according to which the lexical meaning of 'two' is that of the numeral in (26f).

(26f)   <u>Two</u> is an even number.

Together, Polymorphic Comprehension and Substantivalism entail that under-standing any meaning of 'two', including *cardinal* meanings like the one expressed in (26c), requires understanding the meaning of a numeral.

(26c) Mars has two moons.

Thus, given Meaning Comprehension, it follows that one cannot understand 'two' in (26c) without also possessing a concept having the same content as expressed by 'two' in (26f). In other words, understanding cardinal meanings requires possessing a corresponding *numerical* concept.

Jointly, Developmental Priority and Semantic Priority generate a puzzle:

The Developmental Puzzle: How can children understand cardinal meanings of number words, if such understanding requires possessing numerical con-cepts, and yet children only possess numerical concepts *after* possessing cardinal concepts?

In future work, we argue that Strong Diachronic Pluralism provides the best available solution. Roughly put, this involves distinguishing two kinds of count concepts CP-knowers can possess:

Proto-Cardinal Concepts: $\text{TWO}_{proto-card}$, applicable to collections in one-to-one correspondence with sequences of natural language count words, such as ⟨'one', 'two'⟩.

Mature Cardinal Concepts: $\text{TWO}_{mature-card}$, applicable to collections in one-to-one correspondence with the sequence of natural numbers ⟨1, 2⟩.

We argue that early CP-knowers possess *proto*-cardinal concepts, and only later come to possess mature cardinal concepts, once they possess concepts repre-senting numbers as objects, such as $\text{TWO}_{num-ref}$. If so, then the problem with the Developmental Puzzle is Semantic Priority. Specifically, it equivocates: while it's true that possessing *mature* cardinal concepts requires possessing numerical concepts, possessing *proto*-cardinal concepts does not. Nevertheless, possessing either kind of cardinal concept suffices for understanding cardinal uses of number words. Or so we maintain.

4.4 Conclusion

Throughout this Element, we have sought to sketch ways in which fields as diverse as developmental psychology, the philosophy of mathematics, and linguistic semantics might be mutually informative regarding issues about number cognition. We have focused here on three specific points of contact:

the relationship between developmental psychology and foundational theories of arithmetic, the metaphysics of number, and issues regarding conceptual monism and pluralism. Additionally, we suggested further connections in Sections 1.1 and 1.2. We hope that you find these issues as interesting as we do, and that our discussion sparks further interdisciplinary inquiry.

But we are only scratching the surface. Taking seriously the mutual concerns and resources of psychology, philosophy, and linguistics – and endeavoring to bring them into alignment – will likely generate a host of further foci for research. To illustrate, we conclude by posing some further questions arising from the foregoing discussion of Strong Diachronic Pluralism. For heuristic purposes, we divide them into three kinds.

The first falls squarely within the remit of developmental psychology. To begin, note that if Strong Diachronic Pluralism is correct, then Complex Transition follows, since proto-cardinal concepts and mature cardinal concepts are different in kind. Yet this raises a host of issues about which we currently know almost nothing. For instance, if CP-knowers acquire proto-cardinal concepts prior to mature cardinal concepts, how is the transition from the possession of one to the other effected? Specifically, does the acquisition of mature cardinal concepts depend in some important way upon the prior possession of proto-cardinal concepts? If so, then by what means does the child succeed in acquiring mature cardinal concepts from this prior state, given that, by hypothesis, the possession of mature cardinal concepts requires possessing numerical concepts? It would seem that, somehow, proto-cardinal concepts must not merely furnish resources for the acquisition of mature cardinal concepts, but numerical concepts as well. But how? Does it, for example, depend on an enhanced understanding of counting, or does it proceed via some other route? Does it involve a conceptual discontinuity of some sort, or does it require only those conceptual resources previously available to the child? Finally, is the transition rapid and effortless, or is it laborious and hard won in the way acquiring proto-cardinal concepts would appear to be? We simply don't know.

A second family of questions, though relevant to developmental psychology, fall more clearly within the remit of psycholinguistics and, perhaps, the philosophies of mind and language. Broadly speaking, they concern the *conceptual* requirements on word comprehension. Above, we motivated three assumptions jointly implying that understanding cardinal meanings of number words requires possessing numerical concepts. However, in light of those assumptions, Strong Diachronic Pluralism may appear to have puzzling implications. For example, if proto-cardinal and mature cardinal concepts have different contents, then does this imply, perhaps implausibly, a previously unrecognized *ambiguity* in number words? For example, is 'two' ambiguous between a proto-cardinal meaning and a mature cardinal meaning? If so, then what differentiates those meanings, and what independent empirical support, if any, is there for the existence of that ambiguity? Furthermore, if understanding a number word requires possessing a concept having the same content as its lexical meaning, and yet only later CP-knowers possess such a concept, then do early CP-knowers *fail* to understand cardinal meanings of number words, despite their success on various cardinality-related tasks (Section 2.4)? If not, then how is such understanding achievable, given that early CP-knowers cannot yet represent numbers as objects? Again, we think these questions suggest fertile ground for future inquiry.

A final sort of question concerns the metaphysics of number. In Section 4.2 we criticized a range of metaphysical theses found within the Mainstream: term formalism, cultural constructionism, and the Frege–Russell characterization. However, this should not foreclose the possibility that NCR may fruitfully contribute to ongoing metaphysical debates. Specifically, since NCR is centrally concerned with the nature and content of our actual number concepts, it appears well-suited to contributing to a *descriptive* metaphysics of number – an account of what numbers would be like if they were accurately represented by the conceptual scheme we actually have (Section 1.2). Though we are inclined to think that NCR affords many potential resources for metaphysical speculation, as witnessed, e.g. by extant versions of mathematical empiricism (Section 1.2),[35] one avenue we are especially keen to investigate relates directly to Strong Diachronic Pluralism. In particular, in future work we suggest that solving the Developmental Puzzle most plausibly requires adopting a specific metaphysics of the natural numbers: *structuralism* (Section 4.1). Thus, viewed as a descriptive metaphysics of the naturals, structuralism turns out to be a kind of by-product of an empirically plausible account of how later CP-knowers come to represent numbers as objects, and thereby possess numerical concepts.

---

[35] See also Maddy (2018) for related themes.

References

Barwise, J. and Cooper, R. (1981). Generalized quantifiers and natural language. *Linguistics and Philosophy*, 1:413–458.

Beck, J. (2019). Perception is analog: The argument from weber's law. *The Journal of Philosophy*, 116(6):319–349.

Benacerraf, P. (1965). What numbers could not be. *The Philosophical Review*, 74(1):47–73.

Benacerraf, P. (1973). Mathematical truth. *Journal of Philosophy*, 70(19):661–679.

Bermúdez, J. and Cahen, A. (2020). Nonconceptual mental content. In Zalta, E. N., editor, *The Stanford Encyclopedia of Philosophy*. Metaphysics Research Lab, Stanford University, Summer 2020 edition.

Bloom, P. and Wynn, K. (1997). Linguistic cues in the acquisition of number words. *Journal of Child language*, 24(3):511–533.

Burge, T. (2010). *Origins of Objectivity*. Oxford University Press.

Butterworth, B. (2005). The development of arithmetical abilities. *Journal of Child Psychology and Psychiatry*, 46(1):3–18.

Carey, S. (2000). The origin of concepts. *Journal of Cognition and Development*, 1(1):37–41.

Carey, S. (2009a). *The Origin of Concepts*. Oxford University Press.

Carey, S. (2009b). Where our number concepts come from. *The Journal of Philosophy*, 106(4):220–254.

Carey, S. (2011). Précis of the origin of concepts. *Behavioral and Brain Sciences*, 34(3):113–124.

Carey, S. and Barner, D. (2019). Ontogenetic origins of human integer representations. *Trends in Cognitive Sciences*, 23(10):823–835.

Cheung, P., Rubenson, M., and Barner, D. (2017). To infinity and beyond: Children generalize the successor function to all possible numbers years after learning to count. *Cognitive Psychology*, 92:22–36.

Chierchia, G. (1998a). Plurality of mass nouns and the notion of "semantic parameter." In Rothstein, S., ed., *Events and Grammar*. Springer.

Chierchia, G. (1998b). Reference to kinds across languages. *Natural Language Semantics*, 6:339–405.

Chomsky, N. (1987). *Language and Problems of Knowledge the Managua Lectures*. MIT Press.

Chomsky, N. (2014). *The Minimalist Program*. MIT press.

Clarke, S. and Beck, J. (2021). The number sense represents (rational) numbers. *Behavioral and Brain Sciences*, 44:1–33.

Cole, J. (2013). Towards an institutional account of the objectivity, necessity, and atemporality of mathematics. *Philosophia Mathematica*, 21(1):9–36.

Cowie, F. (1998). *What's within? Nativism Reconsidered*. Oxford University Press.

Davidson, K., Eng, K., and Barner, D. (2012). Does learning to count involve a semantic induction? *Cognition*, 123(1):162–173.

Davies, M. (2015). Knowledge–explicit, implicit and tacit: Philosophical aspects. *International Encyclopedia of the Social & Behavioral Sciences*, 13:74–90.

de Cruz, H. (2016). Numerical cognition and mathematical realism. *Philosophers' Imprint*, 16:1–13.

Dehaene, S. (2011). *The Number Sense: How the Mind Creates Mathematics*. Oxford University Press.

Dummett, M. (1991a). *Frege: Philosophy of Mathematics*. Harvard University Press.

Dummett, M. (1991b). *The Logical Basis of Metaphysics*. Harvard University Press.

Egan, F. (2012). Representationalism. In Margolis, E., Samuels, R., and Stich, S., editors, *The Oxford Handbook of Philosophy and Cognitive Science*, pages 249–272. Oxford University Press.

Field, H. (1989). Fictionalism, epistemology, and modality. *Realism, Mathematics and Modality*, pages 1–52. Oxford: Basil Blackwell.

Fodor, J. A. (1975). *The Language of Thought*, volume 5. Harvard University Press.

Fodor, J. A. (1981). The present status of the innateness controversy. In Fodor, J., editor, *RePresentations: Philosophical Essays on the Foundations of Cognitive Science*, pages 257–316. MIT Press.

Fodor, J. A. (1990). A theory of content i. In Fodor, J. A., editor, *A Theory of Content*. MIT Press.

Fodor, J. A. (1992). A theory of the child's theory of mind. *Cognition*, 44:283–296.

Fodor, J. A. (1998). *Concepts: Where Cognitive Science Went Wrong*. Oxford University Press.

Frege, G. (1884). *Grundlagen der Arithmetik*. Wilhelm Koebner.

Frege, G. (1903). *Grundgesetze der Arithmetik II*. Olms.

Fuson, K. C. (2012). *Children's Counting and Concepts of Number*. Springer Science & Business Media.

Gallistel, C., Gelman, R., and Cordes, S. (2006). The cultural and evolutionary history of the real numbers. In S. C. Levinson & P. Jaisson editor, *Evolution and Culture*, page 247. MIT Press.

Gallistel, C. R. (2021). The approximate number system represents magnitude and precision. *Behavioral and Brain Sciences*, 44:e187–e187.

Gallistel, C. R. and Gelman, R. (1992). Preverbal and verbal counting and computation. *Cognition*, 44(1–2):43–74.

Gallistel, C. R. and Gelman, R. (2000). Non-verbal numerical cognition: From reals to integers. *Trends in Cognitive Sciences*, 4(2):59–65.

Geach, P. T. (1972). *Logic Matters*. University of California Press.

Gelman, R. and Gallistel, C. R. (1986). *The Child's Understanding of Number*. Harvard University Press.

Gelman, S. A., Leslie, S.-J., Gelman, R., and Leslie, A. (2019). Do children recall numbers as generic? A strong test of the generics-as-default hypothesis. *Language Learning and Development*, 15(3):217–231.

Goodman, N. and Quine, W. V. O. (1947). Steps toward a constructive nominalism. *Journal of Symbolic Logic*, 12(4):105–122.

Grinstead, J., MacSwan, J., Curtiss, S., and Gelman, R. (1997). The independence of language and number. In *Twenty-Second Boston University Conference on Language Development*.

Hale, B. and Wright, C. (2001). *The Reason's Proper Study: Towards a Neo-Fregean Philosophy of Mathematics*. Oxford University Press.

Harnish, R. M. (2000). *Minds, Brains, Computers: An Historical Introduction to the Foundations of Cognitive Science*. Wiley-Blackwell.

Hart, W. (1991). Benacerraf's dilemma. *Crítica: Revista Hispanoamericana de Filosofía*, 23(68):87–103.

Horn, L. (1972). *On the Semantic Properties of Logical Operators*. PhD thesis, University of California, Los Angeles.

Hurford, J. R. (1987). *Language and Number: The Emergence of a Cognitive System*. Basil Blackwell Oxford.

Hyde, D. C., Simon, C. E., Berteletti, I., and Mou, Y. (2017). The relationship between non-verbal systems of number and counting development: A neural signatures approach. *Developmental Science*, 20(6):e12464.

Kadosh, R. C. and Dowker, A. (2015). *The Oxford Handbook of Numerical Cognition*. Oxford Library of Psychology.

Kennedy, C. and Syrett, K. (2022). Numerals denote degree quantifiers: Evidence from child language. In *Measurements, Numerals and Scales: Essays in Honour of Stephanie Solt*, pages 135–162. Springer.

Kim, J. (1981). The role of perception in a priori knowledge: Some remarks. *Philosophical Studies: An International Journal for Philosophy in the Analytic Tradition*, 40(3):339–354.

Kratzer, A. and Heim, I. (1998). *Semantics in Generative Grammar*, volume 1185. Blackwell Oxford.

Landman, F. (1989). Groups, i. *Linguistics and Philosophy*, 12(5):559–605.

Laurence, S. and Margolis, E. (2005). Number and natural. *The Innate Mind: Structure and Contents*, 1:216.

Laurence, S. and Margolis, E. (2007). Linguistic determinism and the innate basis of number. In *The Innate Mind*, pages 139–169.

Le Corre, M. and Carey, S. (2007). One, two, three, four, nothing more: An investigation of the conceptual sources of the verbal counting principles. *Cognition*, 105(2):395–438.

Lee, A. Y., Myers, J., and Rabin, G. O. (2022). The structure of analog representation. *Noûs*, 57(1):209–237.

Leslie, A. M., Gelman, R., and Gallistel, C. (2008). The generative basis of natural number concepts. *Trends in Cognitive Sciences*, 12(6):213–218.

Leslie, A. M., Xu, F., Tremoulet, P. D., and Scholl, B. J. (1998). Indexing and the object concept: Developing "what" and "where" systems *Trends in Cognitive Sciences*, 2(1):10–18.

Link, G. (1983). The logical analysis of plurals and mass terms: A lattice-theoretic approach. In Bäuerle, R., Schwarze, C., and von Stechow, A., editors, *Meaning, Use, and Interpretation of Language*, pages 303–323. de Gruyter.

Linnebo, Ø. (2009). The individuation of the natural numbers. In Bueno, O. and Linnebo, Ø., editors, *New Waves in Philosophy of Mathematics*, pages 220–238. Palgrave-MacMillan.

Linnebo, Ø. (2017). *Philosophy of Mathematics*. Princeton University Press.

Linnebo, Ø. and Shapiro, S. (2017). Actual and potential infinity. *Noûs*, 53(1):160–191.

Macnamara, J. T. (1986). *A border dispute: the place of logic in psychology*. MIT Press.

Maddy, P. (1990). *Realism in Mathematics*. Oxford University Press.

Maddy, P. (2018). Psychology and the a priori sciences. In Bangu, S., editor, *Naturalizing Logico-Mathematical Knowledge*, pages 15–29. Routledge.

Margolis, E. (2020). The small number system. *Philosophy of Science*, 87(1):113–134.

Margolis, E. and Laurence, S. (2008). How to learn the natural numbers: Inductive inference and the acquisition of number concepts. *Cognition*, 106(2):924–939.

Margolis, E. and Laurence, S. (2013). In defense of nativism. *Philosophical Studies*, 165(2):693–718.

Margolis, E. and Laurence, S. (2022). Concepts. In Zalta, E. N. and Nodelman, U., editors, *The Stanford Encyclopedia of Philosophy*. Metaphysics Research Lab, Stanford University, Fall 2022 edition.

Margolis, E. E. and Laurence, S. E. (1999). *Concepts: Core Readings*. The MIT Press.

McMurray, B. (2007). Defusing the childhood vocabulary explosion. *Science*, 317(5838):631–631.

McNally, L., de Swart, H., Aloni, M. et al. (2011). Inflection and derivation: How adjectives and nouns refer to abstract objects. In *Proceedings of the 18th Amsterdam Colloquium*, pages 425–434. ILLC.

Meck, W. H. and Church, R. M. (1983). A mode control model of counting and timing processes. *Journal of Experimental Psychology: Animal Behavior Processes*, 9(3):320–356.

Mix, K. S. and Sandhofer, C. M. (2007). Do we need a number sense? In Roberts, M. J., editor, *Integrating the Mind: Domain General vs Domain Specific Processes in Higher Cognition*, pages 293–326. Psychology Press.

Moltmann, F. (2013). Reference to numbers in natural language. *Philosophical Studies*, 162:499–536.

O'Shaughnessy, D. M., Gibson, E., and Piantadosi, S. T. (2021). The cultural origins of symbolic number. *Psychological Review*, 129 (6):1442.

Partee, B. (1986). Noun phrase interpretation and type-shifting principles. In Groenendijk, J., de Jongh, D., and Stokhof, M., editors, *Studies in Discourse Representation Theory and the Theory of Generalized Quantifiers*. Foris. pages 115–143.

Piaget, J. and Chomsky, N. (1980). Opening the debate: The psychogenesis of knowledge and its epistemological significance. In Piattelli-Palmarini, M., editor, *Language and learning: The debate between Jean Piaget and Noam Chomsky*, pages 23–34. Cambridge, MA: Harvard University.

Pinker, S. (2007). *The Stuff of Thought: Language as a Window into Human Nature*. Penguin.

Ramsey, W. (2022). Implicit mental representation. In Bangu, S., (editor), *The Routledge Handbook of Philosophy and Implicit Cognition*. Routledge.

Ramsey, W. M. (2007). *Representation Reconsidered*. Cambridge University Press.

Resnik, M. (1997). *Mathematics as a Science of Patterns*. Oxford University Press.

Rips, L. J., Bloomfield, A., and Asmuth, J. (2008). From numerical concepts to concepts of number. *Behavioral and Brain Sciences*, 31(6):623–642.

Russell, B. (1919). *Introduction to Mathematical Philosophy*. Dover.

Sanford, E. M. and Halberda, J. (2023). Successful discrimination of tiny numerical differences. *Journal of Numerical Cognition*, 9(1):196–205.

Sarnecka, B. W. and Carey, S. (2008). How counting represents number: What children must learn and when they learn it. *Cognition*, 108(3):662–674.

Schlimm, D. (2018). Numbers through numerals: The constitutive role of external representations. In *Naturalizing Logico-Mathematical Knowledge*, pages 195–217. Routledge.

Scholl, B. J. and Leslie, A. M. (1999). Explaining the infant's object concept: Beyond the perception/cognition dichotomy. In Lepore, E., and Pylyshyn, Z., (editors), *What Is Cognitive Science*, pages 26–73. Blackwell.

Scontras, G. (2014). *The Semantics of Measurement*. PhD thesis, Harvard University.

Searle, J. R. (1982). The Chinese room revisited. *Behavioral and Brain Sciences*, 5(2):345–348.

Shapiro, S. (1997). *Philosophy of Mathematics: Structure and Ontology*. Oxford University Press.

Shapiro, S. (2000). *Thinking about Mathematics: The Philosophy of Mathematics*. Oxford University Press.

Shea, N. (2018). *Representation in Cognitive Science*. Oxford University Press.

Simon, T. J. (1997). Reconceptualizing the origins of number knowledge: A "non-numerical" account. *Cognitive Development*, 12(3):349–372.

Snyder, E. (2017). Numbers and cardinalities: What's really wrong with the easy argument? *Linguistics and Philosophy*, 40:373–400.

Snyder, E. (2021a). Counting, measuring, and the fractional cardinalities puzzle. *Linguistics and Philosophy*, 44(3):513–550.

Snyder, E. (2021b). *Semantics and the Ontology of Number*. Cambridge University Press.

Snyder, E., Samuels, R., and Shapiro, S. (2018a). Neologicism, Frege's constraint, and the Frege-heck condition. *Noûs*, 54(1):54–77.

Snyder, E., Samuels, R., and Shapiro, S. (2022). Resolving Frege's other puzzle. *Philosophia Mathematica*, 30(1):59–87.

Snyder, E., Samuels, R., and Shaprio, S. (2019). Hale's argument from transitive counting. *Synthese*, 198(3):1905–1933.

Snyder, E. and Shapiro, S. (2022). Groups, sets, and paradox. *Linguistics and Philosophy*, 45(6):1277–1313.

Snyder, E., Shapiro, S., and Samuels, R. (2018b). Cardinals, ordinals, and the prospects for a Fregean foundation. *Royal Institute of Philosophy Supplements*, 82:77–107.

Spelke, E. S. (2000). Core knowledge. *American Psychologist*, 55(11): 1233–1250.

Spelke, E. S. (2003). What makes us smart? Core knowledgeand natural language. In Gentner, D., and Goldin-Meadow, S., editors, *Language in Mind: Advances in the Study of Language and Thought*, pages 277–311. MIT Press.

Spelke, E. S. (2017). Core knowledge, language, and number. *Language Learning and Development*, 13(2):147–170.

Spelke, E. S. (2022). *What Babies Know: Core Knowledge and Composition.* Volume 1. Oxford University Press.

Strawson, P. F. (2002). *Individuals*. Routledge.

Tennant, N. (1997). *The Taming of the True*. Oxford University Press.

vanMarle, K. (2018). What happens when a child learns to count? The development of the number concept. In Bangu, S., editor, *Naturalizing Logico-Mathematical Knowledge: Approaches from Philosophy, Psychology and Cognitive Science*, pages 131–147. Routledge.

Weir, A. (2022). Formalism in the philosophy of Mathematics. In Zalta, E. N., editor, *The Stanford Encyclopedia of Philosophy*. Metaphysics Research Lab, Stanford University.

Wright, C. (2000). Neo-Fregean foundations for real analysis: Some reflections on Frege's constraint. *Notre Dame Journal of Formal Logic*, 41:317–334.

Wynn, K. (1992). Children's acquisition of the number words and the counting system. *Cognitive Psychology*, 24(2):220–251.

Wynn, K. (2018). Origins of numerical knowledge. In Bangu, S., editor, *Naturalizing Logico-Mathematical Knowledge*, pages 106–130. Routledge.

Yi, B.-u. (2018). Numerical cognition and mathematical knowledge: The plural property view. In Bangu, S., editor, *Naturalizing Logico-Mathematical Knowledge*, pages 52–88. Routledge.

Cambridge Elements ☰

The Philosophy of Mathematics

Penelope Rush
University of Tasmania
From the time Penny Rush completed her thesis in the philosophy of mathematics (2005), she has worked continuously on themes around the realism/anti-realism divide and the nature of mathematics. Her edited collection *The Metaphysics of Logic* (Cambridge University Press, 2014), and forthcoming essay 'Metaphysical Optimism' (*Philosophy Supplement*), highlight a particular interest in the idea of reality itself and curiosity and respect as important philosophical methodologies.

Stewart Shapiro
The Ohio State University
Stewart Shapiro is the O'Donnell Professor of Philosophy at The Ohio State University, a Distinguished Visiting Professor at the University of Connecticut, and a Professorial Fellow at the University of Oslo. His major works include *Foundations without Foundationalism* (1991), *Philosophy of Mathematics: Structure and Ontology* (1997), *Vagueness in Context* (2006), and *Varieties of Logic* (2014). He has taught courses in logic, philosophy of mathematics, metaphysics, epistemology, philosophy of religion, Jewish philosophy, social and political philosophy, and medical ethics.

About the Series

This Cambridge Elements series provides an extensive overview of the philosophy of mathematics in its many and varied forms. Distinguished authors will provide an up-to-date summary of the results of current research in their fields and give their own take on what they believe are the most significant debates influencing research, drawing original conclusions.

Cambridge Elements $\equiv$

The Philosophy of Mathematics

Printed in the United States
by Baker & Taylor Publisher Services